Birds & Butterflies of Delhi

Mehran Zaidi

Foreword
Asad R. Rahmani
Director, *Bombay Natural History Society*

Tara Press

Birds & Butterflies of Delhi
Mehran Zaidi

Tara Press
(An Imprint of India Research Press)
Flat No. 6, Khan Market, New Delhi – 110 003
Ph.: 24694610; Fax : 24618637
Website: www.indiaresearchpress.com
E-mail: contact@indiaresearchpress.com; bahrisons@vsnl.com

2010

Illustrations (Birds) Mohammad Anwar
Illustrations (Butterflies) Hans Ram Yadav

ISBN 13 digit: 978-81-8386-055-0
ISBN 10 digit: 81-8386-055-9

Printed for *Tara Press* at Focus Impressions, New Delhi.

Foreword

Birds and butterflies are some of the most loved and admired taxa of animals, due to their beauty, colour, vivacity, diversity, fragility, accessibility and friendliness. Another reason for our admiration for birds and butterflies is their flying ability, which Man has always aspired to gain. Although we have reached the moon, we can only glide (that also with artificial 'wings'), much like the ancestors of birds, millions of years ago. Which child does not desire to majestically soar like an eagle, high up in the sky, or dexterously flit from flower to flower, sipping the invigorating nectar!

Birds and butterflies are also good indicators of the quality of our environment – more of them means that the environment is still good, the air is breathable, trees and shrubs are surviving, open space has not been encroached upon, the wetlands are still not too polluted, and the crow and the cat population are under control. In UK, one of the parameters of the quality of life is the diversity of birds found around a house, a village or a town.

Being more or less harmless to us and found everywhere, bird and butterfly watching can be a very enjoyable hobby. Although, there is no age bar in developing this hobby, the best is to start early. A child-like inquisitiveness will teach you the difference between an Ashy Prinia and the Jungle Prinia, the difference between the call of the Rock Bush Quail and the Jungle Bush Quail, and the subtle plumage difference between the Tawny Eagle and the Greater Spotted Eagle.

This is Mehran Zaidi's second book. I liked his first book *Bird by Bird: Common Indian Birds*. Certainly, a great achievement for a graduate student! I hope, like his earlier book, this book will also help in developing interest among youngsters, especially in environment protection in general, and birds and butterflies in particular. Can a person who enjoys the antics of a Tailor Bird or a Purple Sunbird in his garden be oblivious to the filth which we generate, can a person who spends a lovely morning near a wetland while ducks fly all around him/her, remain unmoved by the pollution of the Yamuna, can a person who likes the Plain Tiger or a Danaid Eggfly allow unnecessary cutting of shrubs in his municipal garden?

Love of nature starts from knowing the life forms around you. And if you love nature, you would protect it. I hope this book would help in developing a young army of conservationists, which our country urgently requires.

Asad R. Rahmani
Director
Bombay Natural History Society

Introduction

Watching the winged creatures in Delhi

Delhi is a haven–or should I say heaven–for birdwatchers. This green city of India has close to 450 species of birds, making it only second to Nairobi amongst the 'bird capitals' of the world. There are many sites in Delhi where one can watch birds. Probably the best place for birding in Delhi is the Yamuna. The stretch from Wazirabad up to Okhla is an ideal birding area. In South Delhi, there is the 'Okhla Bird Park' (OBP), where one can watch a lot of birds, especially during the winter season when the migratory 'guests' arrive. The Yamuna Bio-diversity Park in Wazirabad is an excellent place for birdwatching. An artificial wetland has been created here which attracts water birds in huge numbers. YBP is a must for bird lovers. The Okhla Bird Park and the Yamuna Bio diversity Park are also very good places to spot rare butterflies.

The *Garden of Five Senses* near the Qutub Minar has a butterfly park where one can spot many different species. A 'garden' to please one's ophthalmic sense for sure!

The *Asola Bhatti* Wildlife Sanctuary, located on the Southern ridge in Tughlaqabad, is a very good scrubland forest for watching birds. This Sanctuary is also a good place to spot butterflies.

The *Sultanpur* National Park, near Gurgaon and the *Mohammadabad* marshes, near Sonipat are fine birding spots in the NCR. Although Sultanpur has lost much of its charm, it still is a good birdwatching place and is regularly frequented by birders. That's a lot to begin birding Friends, so go for it. May our tribe increase!

How it all started

The colours mesmerized me and when it suddenly took off, oh so gracefully, its flight made me forget my zipping dinky cars. In excitement I clutched at my *chacha's* neck and repeated what he had said, "Tree-Pie! Tree Pie!" he put me down and the two of us walked on in Buddha Jayanti Park in search of another shy bird. I was five.

For a little boy an Uncle who was a surgeon, a Captain in the army and the captain of the Services Ranji cricket team was THE IDOL. I had to do everything Chacha did. That's how I got hooked on to birds. When other kids my age would rattle off names of 'Gi-Joes' I would have a longer list of birds! As I grew my obsession with birds grew. My parents encouraged me and shared my hobby. By the time I was twelve I was a committed BIRDER. A visit to Bharatpur was our annual pilgrimage. Gradually I became concerned about conservation issues and started writing and involving myself in activities aimed at such concerns.

My fascination for nature's most beautiful creatures, butterflies, developed much later. I am eternally beholden to William Wordsworth for it was a poem of his which drew my attention to these gentle beauties. I quote;

I've watched you now a full half-hour
Self-poised upon that yellow flower

And little Butterfly! Indeed
I know not if you sleep or feed.
Come often to us, fear no wrong
Sit near us on the bough!
We'll talk of sunshine and of song
And summer days, when we were young;

My first chance to share my passion with others came my way when *The Tribune* accepted some write-ups I did on birds. The thrill of seeing my first piece in print remains unbeaten. Around this time I met Ms Arundhati Deosthale who reposed her faith in an eighteen year old and entrusted me with my first book which later was taken up by Ms Sayoni Basu and Scholastic did a beautiful job of it. Thanks Arundhati Aunty, Sayoni Ma'am and Anoushka Ma'am! In the meanwhile I had started doing a column with the *Hindustan Times*, titled SPOTTED IN MY NEIGHBOURHOOD. In this column I wrote about both birds and butterflies, here I must acknowledge Ms Smitha Vijay, Vipul (Mudgal) Sir and Himanshu (Joshi) Sir, who all encouraged me a lot.

However, it is Mr Anuj Bahri, Anuj Uncle, who has introduced me to the world of writing books for grown ups, and finally I am here doing this intro! Anuj Uncle, I truly appreciate the love and care with which you are doing my book.

So, let's begin with some tips for birders.

Birding Etiquette

- **Wear dull Clothes** – Bright and colourful clothes can unsettle or scare away birds.

- **Wear strong and comfortable shoes** – For birdwatching you will have to walk around on rough terrain, or on swampy and marshy grounds. Therefore, a good pair of shoes is very important.
- **Carry binoculars** – Birds are shy, so the best way to watch these beautiful creatures is from a distance and through binoculars.
- **Maintain a record** – A very important aspect of birdwatching is keeping a record of the birds you sight. The name of the bird, the date and the place should be noted. You should also maintain the pictures and videos of the birds in electronic files. This way you can share them with other birdwatchers and friends.
- **Go birding with an experienced birdwatcher** – In your first few birdwatching trips, take an expert with you. He/She will help you find rare birds and the best locations.
- **Take a 'bird book' with you** – This is one of the most important tips for amateurs. *The Book of Indian Birds*, by Salim Ali, is obviously the most preferred. Other useful books are *Birds of India* by Martin Woodcock and *Common Birds of India* by Dr. Asad. R. Rehmani. For specifically birding in Delhi, Ranjit Lal's *Birds of Delhi*, and the *Atlas of the Birds of Delhi* by Nikhil Devesar, Bikram Grewal and Bill Harvey are the best.
- **Be silent and don't move** – Noise and quick movements make the birds nervous and they eventually fly away. So, being still and quiet will help you watch them and study them in detail.
- **Get a local guide** – If you are in a Sanctuary or a National Park get a local guide. The guides know the areas inside out and also know exactly which bird is found where.

- **Join a birdwatching group** – The best way to learn birdwatching quickly is doing it in a group. You get to learn a lot from the expert birdwatchers. E-groups like Delhibird are very active.

I sign off with what I consider an anthem for bird lovers, composed by William Blake;

A Robin Redbreast in a cage
Puts all heaven in a rage.
A Skylark wounded on the wing
Doth make a cherub cease to sing.
He who shall hurt the little wren
Shall never be beloved by men.

Happy Birding!

Mehran Zaidi
New Delhi, September 2009

About the Author

Mehran Zaidi : The youngest *Ornithology* enthusiast, Mehran started watching birds when he was just 10 years old. At 17 years he was a contributor to various National dailies on the subject of Birds and Birdwatching. By the age of 19 he was already an author. Educated at the Modern School, New Delhi and topped his batch in history, he is presently studying at the Asian College of Journalism, Chennai.

Mehran has done a one-year diploma course for *Leadership in Biodiversity Conservation* from the Bombay Natural History Society (BNHS), and was one of the few, besides being the youngest to be awarded a certificate. He is presently undergoing a one-year Ornithology course from BNHS. He is also a keen photographer and has completed a course in Advanced Photography from Bal Bhavan, New Delhi.

Mehran has interned with the Hindustan Times NEXT to upgrade his journalistic skills. He was a regular contributor one 'birds and other creatures' for *The Hindustan Times* and *The Tribune*. He wrote a weekly column, 'Spotted in my Neighbourhood' for HT NEXT which was later moved to the main paper.

The Limca Book of Records acknowledges him as a contributor to the Nature and Transport section of their yearbook.

Mehran regularly undertakes interactive sessions and presentations in schools on nature, conservation and birdwatching.

He is actively involved in the Asola Wildlife Sanctuary at Tughlaqabad and 'Delhibird', with groups of birdwatchers and conservationists.

His first book, on birdwatching, *Bird By Bird* (Scholastic India 2006) has received much critical acclaim. It has been extensively reviewed and appreciated, including a review from the celebrated writer, Khushwant Singh.

Birds and Butterflies of Delhi is his second book on the subject.

Mohd Anwar is a BFA from the prestigious B K College of Art & Craft, Bhubaneswar. His works have been exhibited in the Lokayat Art Gallery and Just Art in 2002 & 2004 respectively. He has illustrated books for Ratna Sagar, Penguin and Scholastic. He specializes in nature, water colours being his favourite medium. He also does murals and collages.

Hans Ram Yadav is a graduate from College of Art, Delhi University and an MA in drawing and painting from Jiwaji University, Gwalior. Yadav illustrates for Ratna Sagar and Rupa and for various magazines and newspapers. He loves to share his talent and knowledge and hence teaches in a government school. He specializes in landscapes and portraits.

Publisher's Note

I have always enjoyed working with the young and first-time writers, who in spite of their age and lack of popularity have shown great talent and professionalism in their respective fields of work and hobbies. They always bring a new and fresh perspective to any topic.

Mehran, is one such enthusiast who has a vast amount of knowledge in the field of Ornithology (study of birds). What he lacks in experience of his technical qualification (because of his age), he makes all efforts to compensate for with his in-depth research as a field observer. He considers Birds and Butterflies his friends and tries to understand their behaviour patterns as a non-technical observer. He tries to explain their place in nature and life around us – in simple words that other birdwatchers and nature lovers find easy to understand.

His new book, *Birds & Butterflies of Delhi* is one such creation targeted at the vast general population that notices all these different variety of birds and butterflies in the areas of Delhi and its environs but do not have the time to follow their pattern and natural cycle through the year. Mehran's new book provides enough information for the first time observer to identify and understand the birds and butterflies around them.

We have tried to keep the detailing simple by showing the observation grounds (outline map of Delhi) where these variety of birds can be found through areas highlighted on the maps along with every variety of bird given. However a general map of the

Delhi area (see Figure 1) that identifies the North/South/East/West break-up of the region is given on this page as a guide.

The book is simply divided into three main sections:

Non-water Birds (■■ section)
Water Birds (■■ section)
Butterflies (■■ section)

The Birds are further classified into three sub-categories:

Resident Birds (■■ spot markings on the map)
Summer Migratory Birds (■■ spot markings on the map)
Winter Migratory Birds (■■ spot markings on the map)

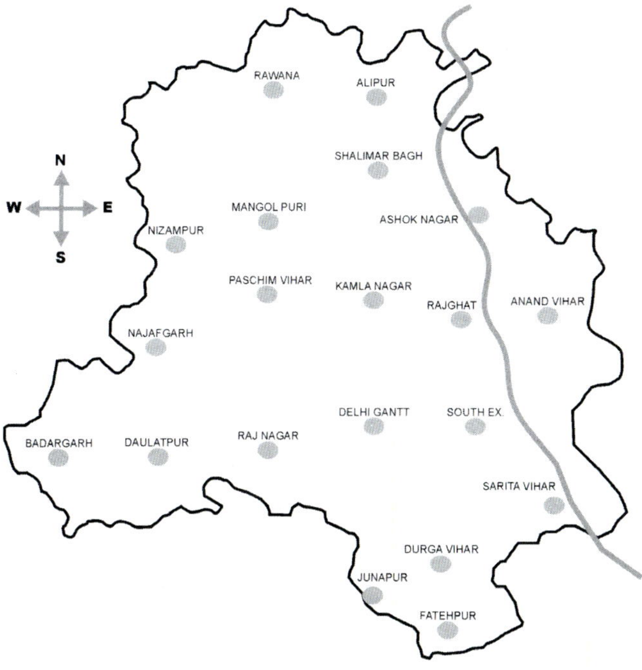

List of Birds and Butterflies

Non-Water Birds (section)

Butterflies (■■ section)

Birds

Ashy Parinia

House Sparrow

Purple Sunbird

House Sparrow

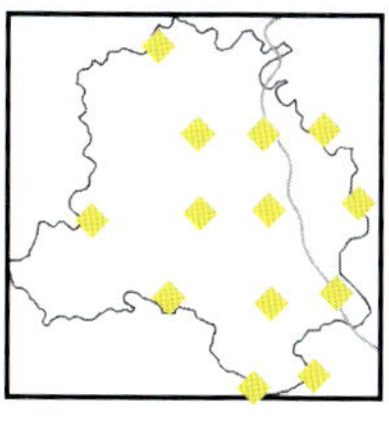

Resident Bird

Size - *15 cm*

Hindi name - *Goriya*

Scientific name - *Passer domesticus*

We all know our little friend the House Sparrow. It is a greyish-brown and black bird with a white patch on the underside of its tail. The male sparrow has darker colours.

Sparrows are found everywhere. At home, in the market, in the office, a sparrow is the most familiar bird. It is found anywhere where humans are found. We can see it bathing and playing around in water puddles or simply flitting around chirruping continuously.

Dozens of sparrows roost in trees, when dusk falls. They get noisy while roosting and create quite a ruckus.

The nesting season of the House Sparrow is almost throughout the year. Its nest is a collection of straw, feathers and man-made stuff like thread, plastic pieces, paper, etc. It lays 3 to 5 eggs which are a pale greenish-white, and are blotched with brown.

For the uninitiated locals in North India the Sparrow is referred to as 'Chirlya', meaning bird!

Food - Grass, cereal grains, fruit buds, and insects. City sparrows largely depend on humans for food.

Call - Chirruping *cheer-cheer*, etc.

Purple Sunbird

Ashy Prinia

Oriental White-Eye

Ashy Prinia

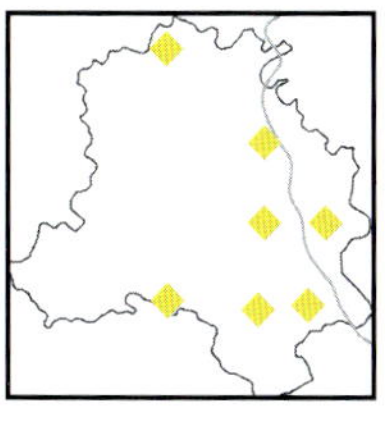

Resident Bird

Size - *13 cm*

Hindi name - *Phutki*

Scientific name - *Prinia socialis*

The Ashy Prinia, is a very small and cute bird which I regularly see near my house in North Delhi, mostly in the garden. It is one of our most cheerful garden birds. This prinia is a resident breeder in the Indian subcontinent.

The Ashy Prinia has short rounded wings and a long tail which frequently becomes erect as it is shaken up and down. It has well-built legs and a little black bill. In breeding plumage, adults are ashy grey above, and have a black eye band. The underparts are whitish but yellow on the sides. The sexes are alike except that the male has a darker bill in the breeding season. The Ashy Prinia has a weak and uncertain flight. This lovable little bird is mostly found in gardens, parks, dry open grassland, and also open woodlands.

The Ashy Prinia lays 3 to 4 eggs. Tree-pies, cats, crows, and most of all, the frequent use of pesticides by humans, are the major threats to the eggs and babies of the Ashy Prinia.

Food - A wide range of insects and spiders and also insect-eggs
Call - Sweet and buzzing, a repetitive *tchup, tchup, tchup*

Oriental White-Eye

Purple Sunbird

Red-Vented Bulbul

Purple Sunbird

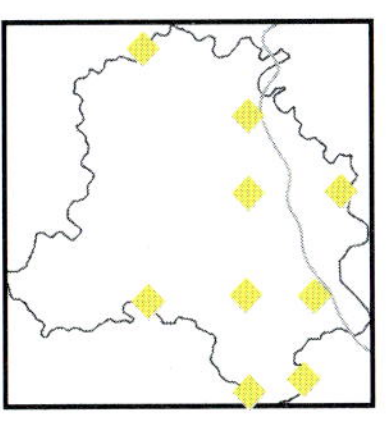

Hindi name - *Phool soongni*

Scientific name - *Nectarinia asiatica*

Resident Bird

Purple Sunbirds are very petite and elegant birds. The adult male is a pretty glossy purple. The female has yellow-grey upperparts and yellowish underparts. They have thin down-curved bills.

This species is found in a variety of habitats with some trees, including forests, gardens and cultivation. It is quite widespread in the Delhi region, where you can spot it in any large park, like the Buddha Jayanti Park.

The nesting season of the Purple Sunbird is mostly March to May. The nest is a type of an oblong pouch of grass, cobwebs, etc. It usually hangs from the tip of a branch of a bush or a creeper. It lays 2 to 3, greyish-white eggs marked with brown. The female builds the nest and incubates the eggs. The male helps in feeding the young.

Food - Insects, spiders and flower nectar
Call - The call is a humming *cheeit- cheeit*

Red-Vented Bulbul

Oriental White-Eye

Baya Weaver

Oriental White-Eye

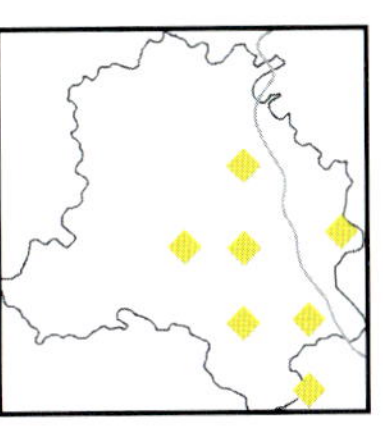

Resident Bird

Size - *10 cm*

Hindi name - *Baboona*

Scientific name - *Zosterops palpebrosus*

Have you ever seen a bird wearing spectacles? Well, the Oriental White-eye does look as if it is wearing big white-rimmed spectacles! The White-eye is an adorable diminutive bird smaller even than the sparrow. It has very noticeable white circles around its eyes. The whole upper plumage and the neck is greenish yellow whereas the abdomen is whitish grey. The sexes are exactly alike in appearance.

The Oriental White-eye is a familiar arboreal resident of Delhi's gardens, groves and parks. The Buddha Jayanti Park is a good place to check them out.

The White-eye is found in groups of up to forty birds. That makes for a pretty-looking colony! It limits itself generally to blossoming branches and bushes, and does not like to venture out in the open. That's why its hard to spot, and, like the Koel, more often heard than seen.

The Oriental White-eye breeds between February and just before the start of the rainy season in June-July. This is to protect the fragile little nest from the heavy rains. It lays 2 to 3, small pale blue eggs. Both parents build the tiny nest together and share the 'domestic' duties. The next time you see the Oriental White-eye give it the 'glad eye', as it's a rare sight!

Food - Mainly insects, flower nectar and berries
Call - A soft and melancholic *tseer... tseer*

Baya Weaver

Red-Vented Bulbul

Common Myna

Red-Vented Bulbul

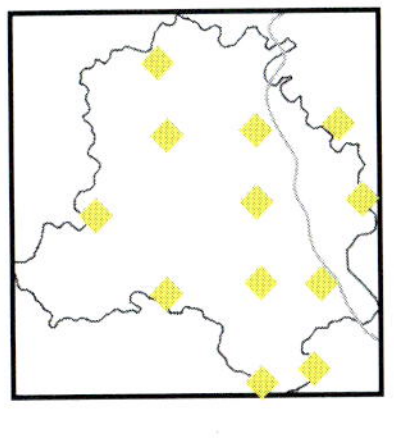
Resident Bird

Size - *20 cm*

Hindi name - *Bulbul*

Scientific name - *Pycnonotus cafer*

You must have noticed a graceful, crested-brown bird hopping around in parks and gardens. This playful and charming bird is the Red-vented Bulbul, the legendary bird of many stories and the beloved of the rose in Urdu poetry.

It is a fine-looking brown bird with a black crested-head and a red patch under the root of its tail. The sexes are alike in plumage, but juvenile birds are duller than the adults. The flight of the Red-vented Bulbul is lively and similar to a woodpecker's.

The Red-vented Bulbul is without doubt one of the most easily sighted birds in the Capital. It is found in gardens, city parks and shrubberies, especially on the northern ridge. Many a time I have spotted them merrily playing around at the Mahavir Vanasthali.

Belying their graceful and friendly persona, Red-vented Bulbuls are belligerent warriors. So they are tamed and kept as fighting pets. The bulbul is very affectionate towards its mate. You will mostly spot two bulbuls sitting close to each other. These birds are gregarious and have a penchant for flying in pairs or flocks during the non-breeding season.

The nesting season of the Red-vented bulbul is from February to May. Its nest is cup-shaped, and made from small roots, in a tree or in a bush. It lays 2 to 3 pinkish-white eggs, blemished with purplish-brown. Both parents share the domestic duties.

The bulbul is sometimes erroneously referred to as the nightingale, so much so that Sarojini Naidu known as the Nightingale of India in English, was called Bulbul-e-Hind in Urdu!

Food - Berries, seeds, nectar, flower buds, ants and other small insects. It doesn't mind a peck at your food too!

Call - A musical and invigorating *peep-peep*.

Common Myna

Baya Weaver

Brahminy Starling

Baya Weaver

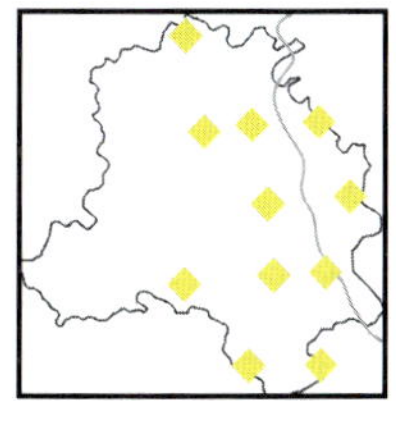

Resident Bird

Size - *15 cm*

Hindi name - *Baya*

Scientific name - *Ploceus philippinus*

The Baya Weaver is one of the most fascinating birds as it's a master craftsman. The male is dark brown above with yellow streaks, a dark brown neck and a yellow breast. It has a yellow crown. The female is buff above with darker streaks. The Baya has a strong conical bill. Its bill is its high point, literally and figuratively, as we shall find out soon.

The Baya Weaver is a gregarious bird found in big flocks in open cultivation like paddy and other grain-harvested fields. It can damage ripening crops. In Delhi, you can find these skilful birds near the Okhla Barrage and on the ridge.

The Baya, likes to build its nest usually over or near water. It is called a 'weaver' because it weaves an amazing home for itself. The nesting season is between May and September. The nest is a vase-shaped structure, made out of paddy leaves and grass, suspended in clusters from twigs, mostly over water. Pieces of mud are plastered inside the nest near the eggs. The Baya lays 2 to 4, snowy-white eggs. The male builds the nest by itself, while the duty of incubation is the female's alone. Each male has many nests and many females at the same time. That's a casanova, if ever there was one!

Food - Cereals and sometimes insects
Call - A *chit-chit-chit* sound

Brahminy Starling

Common Myna

Magpie Robin

Common Myna

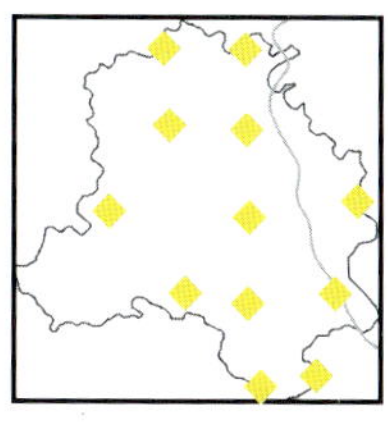

Resident Bird

Size - *23 cm*

Hindi name - *Desi myna*

Scientific name - *Acridotheres tristis*

There are many types of mynas found in Delhi, like the Brahminy Myna, the Bank Myna and the Pied Myna, but the most common is of course the Common Myna!

The Common Myna has a brown body and wing plumage, with large white wing patches visible in flight. The head and the throat are dark grey. The bill, the bare skin around the eyes and the strong legs are bright yellow. The sexes look similar. It walks with occasional hops on the ground. The Common Myna is typically found in open woodland, cultivation and mostly around human habitation.

The nesting season of the Common Myna is from April to August. Its nest is a collection of roots, twigs and other man-made materials like thread, paper, etc. It lays 4 to 5, glossy-blue eggs. Both sexes share the domestic duties.

The myna looks at you with a sharp eye! It is a very intelligent bird, and is also quiet genial. This means it likes to mix around with friends.

Regional poetry and literature as well as Hindi film songs speak of the love stories of 'tota-myna'- the Parakeet and the Myna! Wonder why!?

Food - Insects, fruits and living in the company of humans, food stolen from households. You can also find mynas at garbage dumps searching for food. They forage on the ground for insects, especially grasshoppers.

Call - A variety of croaks, squawks, chirps, clicks and whistles

Magpie Robin

Brahminy Starling

Indian Robin

Brahminy Starling

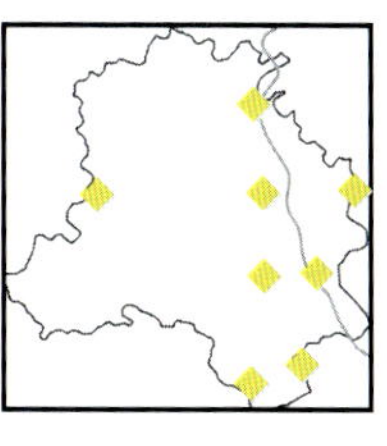

Resident Bird

Size - *22 cm*

Hindi name - *Brahminy myna*

Scientific name - *Strunus pagodarum*

The Brahminy Starling is also known as the Brahminy Myna. It is a dignified-looking bird. It is grey above, buff below, and has a black crown, just like a Brahmin's ponytail! The sexes are alike in appearance. The juveniles are almost uniformly dull coloured.

The Brahminy Starling is a common dweller of Delhi's parks. It easily enters gardens and residential compounds. However, it prefers damp grasslands, and spends more time on the ground than on trees. It collects in small groups, but seldom is the flock larger than twenty birds.

The nesting season of the Brahminy Starling is from March to June. Its nest is a mixed assortment of grass, twigs and small roots. It lays 3 to 5 pale blue, unmarked eggs. Both the parents share the domestic duties, but the female is believed to incubate alone.

These days more and more Brahmins are giving up there ponytails, so isn't it time now to think of another name for the Brahminy Starling! May be the Nerd Starling. (In honour of the guys who've borrowed the ponytail!)

Food - Fruits and berries; grasshoppers, moths, ants, etc
Call - Many comical and noisy chattering sounds

Indian Robin

Magpie Robin

Blue Rock Pigeon

Magpie Robin

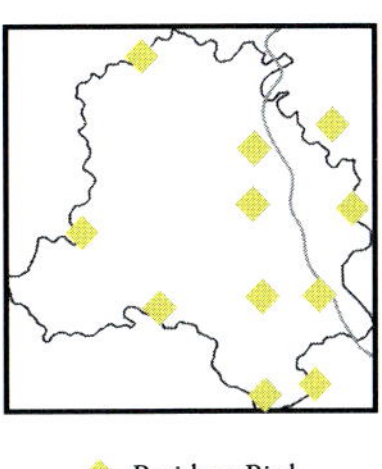

Size - *20 cm*

Hindi name - *Daiyar*

Scientific name - *Copsychus saularis*

Most mornings, in the District Park, Rohini, where I live, I meet this tiny little gentleman 'dressed' in a black tailor coat and a white shirt. It's actually the Magpie Robin! It is one of the more familiar birds in the parks of Delhi.

The male is a small and smart-looking black and white bird. The female is brown and grey in colour. Its tail is raised up. The Magpie Robin is usually seen alone or in pairs, often with other birds in diverse groups. It is a sociable creature. You will find this pretty little bird most active around twilight. It hops around, chirrups away and calls to its mate.

Magpie Robins breed from January to June. They build their nests almost anywhere from thick undergrowth, in branches of small trees, palms, hollows in old trees and even near human habitation, like in a veranda, in a hole in the wall, in an old tin can, and in stables. Nests are usually built low. These nests are big, messy, shallow cups shabbily put together from grass, dried leaves, twigs, moss, and roots. They lay 3-5 eggs, which are pale blue or greenish with brown or purple spots. While the female incubates the eggs, both parents share the responsibility of raising the young ones. A lesson to learn for many human fathers!

Food - A large variety of insects like moths, ants, beetles, caterpillars, etc

Call - The Magpie Robin is one of our best songsters and has a very rich voice. Its call is a beautiful and clear whistle. It sings with its wings partly loose and tail down.

Blue Rock Pigeon

Indian Robin

Spotted Dove

Indian Robin

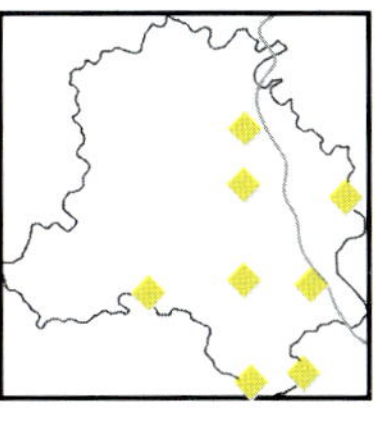

Size - *16 cm*

Hindi name - *Kalchuri*

Scientific name - *Saxicoloides fulicata*

The Indian Robin is an energetic little bird. It is black all over with a white patch on the wing. The root of the cocked-up tail is brick-red in colour. The female is a dull brown without any wing patch.

The Indian Robin is found in pairs in lightly-wooded areas, scrubland, etc. It is generally found near human habitation. It is a shy and confiding bird, and in the Delhi region, you can spot it in any of the large gardens and lawns.

The nesting season of the Indian Robin is from April to June. The nest is a pad of grass, roots, feathers, hair, etc. The nest is made under a stone, the stump of a tree, or even inside a tin can or an earthen pot! The Indian Robin lays 2 to 3 eggs. The eggs are creamish-white in colour, sometimes speckled with brown. The incubation is done by the female but the male helps in the other domestic duties.

My first introduction to the word robin is a vague childhood memory of a blue packet in my granny's house which was called 'Robin' something, and had the picture of a cute little bird. Much later I came to know that the packet contained ultramarine and was a clothes whitener.

Food - Insects and their eggs, spiders, etc
Call - A long *weeech...weeech...* sound

Spotted Dove

Blue Rock Pigeon

Ring Dove

Blue Rock Pigeon

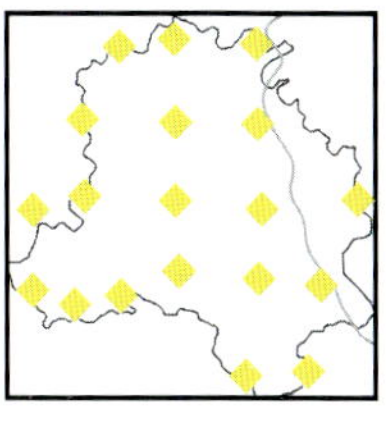

Resident Bird

Size - *33 cm*

Hindi name - *Kabutar*

Scientific name - *Columba livia*

The Blue Rock Pigeon are sturdy birds and can fly high-speed over long distances. It is a familiar grey bird, you must have met often. It has a shining greenish-purple patch on its neck and on the upper part of its breast, hence its name. Pigeons are peaceful and pleasant birds.

Blue Rock Pigeons inhabit ancient monuments and abandoned buildings. In many temples and mosques they are regularly fed by the devotees. You can find huge numbers near the traffic island on Panchkuian Road. It is our very own Trafalgar Square!

The nesting season of the Blue Rock pigeon is undefined. They nest probably all year round. The nest is a fragile collection of sticks on a cliff or a fissure, or in ceilings of old houses and on air-conditoners and desert coolers. The Blue Rock Pigeon lays 2 white eggs. Both the female and the male share the domestic duties.

Long ago pigeons were trained and used as a 'courier service' and were called carrier pigeons.

Food - Grains like maize, and other cereals, pulses, groundnuts, weed seeds, etc.
Call - An interesting and deep *gootrgoo-gootrgoo.*

Ring Dove

Spotted Dove

Rose-Ringed Parakeet

Spotted Dove

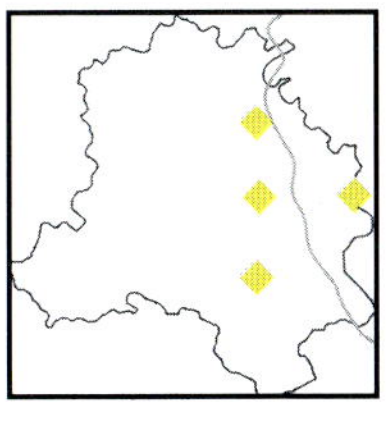

Resident Bird

Size - *30 cm*

Hindi name - *Chitta fakhta*

Scientific name - *Streptopelia chinensis*

The Spotted Dove is a long-tailed, clean-looking bird. Its back, wings and tail are dull brown and spotted with buff. The head and underparts are pinkish, graduating to pale grey on the face and lower belly. There is a black neck patch spotted with white. The legs are red.

The Spotted Dove is a common and widespread species in the open woodlands, farmland and habitations of Delhi and its outskirts. If not disturbed, it even ventures into verandas and gardens of our houses.

The Spotted Dove is largely terrestrial, foraging on the ground in grasslands and cultivation. It is not particularly gregarious, and is usually found alone, or in pairs. It is tame but sudden noises startle it into flight.

Apparently there is no specific nesting season of the Spotted Dove. It may breed any time of the year. The nest is a weak structure of sticks low down on a bush or in a tree. The Spotted Dove lays 2 white eggs. Both the sexes share the domestic duties.

The Dove is used as a symbol of peace, though I have not yet been able to figure out which Dove is shown with the olive branch. To me the bird holding the branch looks like a cross between a dove and a pigeon!

Food - Grass seeds, grains etc.
Call - A low and gentle *coo-coo-croo.*

Rose-Ringed Parakeet

Ring Dove

House Swift

Ring Dove

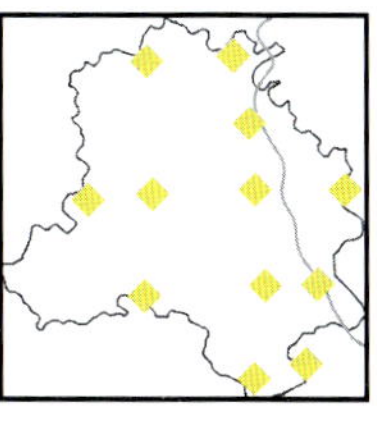

Size - *32 cm*

Hindi name - *Fakhta*

Scientific name - *Streptopelia decaocto*

Living in one of the greener parts of Delhi has its advantages. My apartment is exactly opposite the District Park. So, I am one of the blessed few who wake up to the peaceful and deep song of the Ring Dove calling out, Kukkoo... kook... kukkoo... kok. The sound fills me with contentment. The Dove is considered to be a very serene and gentle bird.

The Ring Dove is also called the Eurasian-collared Dove. It is so named because it has a thin black half-collar on the back of its neck. It is dull grey and brown with a light pink breast.. The sexes look alike. The Ring dove is a very familiar resident bird, found all over Delhi. It is found in deciduous and scrub wooded areas, parks, villages and agricultural cultivation. It prefers trees like the *Dhak* and *Keekar*. It is mostly spotted single or in small parties. It mixes around with other doves.

The Ring Dove easily enters gardens and verandas in search of food.

The nesting season of the Ring Dove is just about all year. The nest is a sparse platform of twigs in a bush or a tree. The Ring Dove lays 2 eggs. Both sexes share the domestic duties.

Mubashir Hasan in his book, Birds of the Indus, quotes a verse of the great Sufi saint and poet Bulleh Shah, who expresses the meekness of the ruler of the day:

Baz chhad gai alna, ghuggi pardhan hoi, (The eagle has discarded its place and the dove has taken charge).

Whenever I see or hear this gentle bird, I am reminded of my primary school grammar lesson on Similies: as gentle as a dove.

Food - Grains and seeds, it prefers to feed on the ground
Call - A peaceful *Kukkoo...kook...kukkoo...kok*

House Swift

Rose-Ringed Parakeet

Hoopoe

Rose-Ringed Parakeet

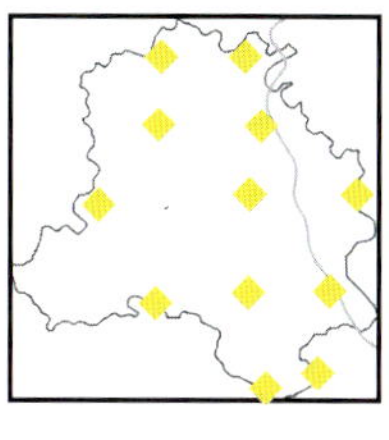

Resident Bird

Hindi name - *Tota*

Scientific name - *Psittacula krameri*

Almost anywhere you go in verdant Delhi, dawn or dusk, you are sure to hear a band of parakeets creating their own screechy music. The rose-ringed Parakeet is the most easily sighted of the parakeets found in Delhi.

The plumage of the Rose-ringed Parakeet is bright green. It has a small, curved, crimson beak. The male has a reddish-pink and black collar around its neck. The female however is plain green.

The Rose-ringed Parakeet is a raucous bird and nests in hollow tree trunks and in crevices and holes of old buildings and monuments. You are sure to come upon many 'happy families' cosily nesting in the Old Fort near the Delhi Zoo, the tall trees in the Pusa Complex and the Tughlaqabad Fort to name a few.

These Parakeets often flock together and attack crops and fruit orchards. A huge flock can totally destroy a standing field. That is why farmers consider them as pests.

These ill-fated birds are caged, sold and kept as pets because they can 'talk'. This means that they can mimic sounds they hear. What a price to pay for a God-given gift!

The nesting season of the Rose-ringed Parakeet is chiefly from February to April. It lays 4 to 6 eggs, which are pearly white in colour.

When I was three, my granny bought me my first alphabet book, which told me that P is for PARROT and carried a beautiful picture of the Rose-ringed Parakeet! It was a decade later when I realized that in 'Indian English', parakeets are called parrots!

Food - Most fruits, crops, cereals, flower nectar
Call - Its call is piercing and loud *eaak... eaak.*

HOOPOE

HOUSE SWIFT

JUNGLE BABBLER

House Swift

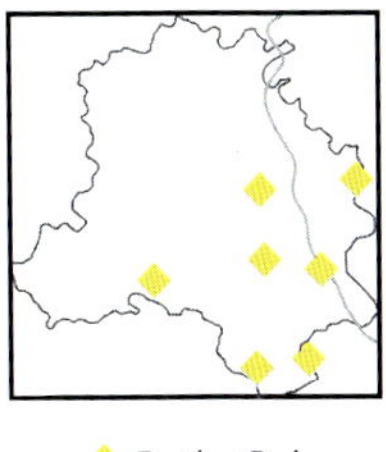

Resident Bird

Size - *15 cm*

Hindi name - *Ababeel*

Scientific name - *Apus affinis*

The House Swift is a petite bird. It is greyish-black, with a white throat, a white rump, a short squarish-tail and long narrow wings. The sexes are alike in appearance.

The House Swift is generally seen close to human habitation, flying over buildings, houses, etc. In Delhi, House Swifts can be spotted at most of the old monuments and buildings. The Swift flies at fantastic speeds, diving and dipping in the sky, chasing insects like flies. Due to a strange foot structure the House Swift cannot perch but can only cling to rough surfaces.

The nesting season of the House Swift is from February and September. The nest is a round shabby cup, made of feathers and straw cemented with the bird's saliva. The nest is located in the ceilings of walls and verandas of buildings. It lays 2 to 4, milky-white eggs. Both sexes share the domestic duties.

The House Swift can't sit. I wonder how it incubates!?

Food - Flying insects
Call - Shrill screams

Jungle Babbler

Hoopoe

Coppersmith Barbet

Hoopoe

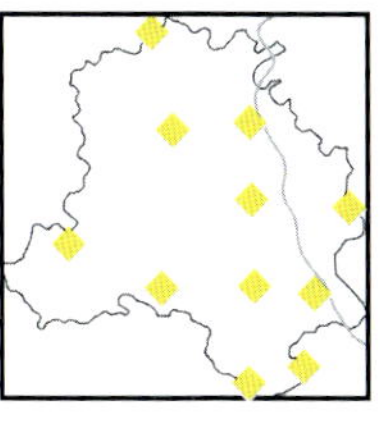

Size - *31 cm*

Hindi name - *Hudhud*

Scientific name - *Upupa epops*

The Hoopoe is a 'legendary' bird. In classical Chinese poetry, the Hoopoe is shown as a messenger of the gods often bringing news of the spring. The Hoopoe is considered very auspicious in China, mainly because of its sheer beauty.

The Hoopoe is an elegant-looking bird with an orangish-brown upper body. The wings, back and tail are black and white in colour. It has a black and white tipped beautiful crown and a long bent-down beak.

The Hoopoe is found throughout the Indian subcontinent, and is mostly seen alone or in scattered pairs. It is a familiar bird in the city's parks, though it is not as frequently seen as it used to be a decade ago, probably due to urbanization and senseless construction.

The nesting period of the Hoopoe is chiefly from February to May. The nest is a natural tree-hollow or a hole in a wall or the ceiling of a building, sloppily lined with rags and straw. The nest is infamous for being fowl-smelling and dirty. The Hoopoe lays 5 to 6 white eggs.

A legend goes that when people were mercilessly killing hoopoes, enamoured of their golden crest, the Hoopoe begged King Solomon to save its race from dying out. The great king was so moved by its plight, that he bestowed it with a black-tipped orangish-brown crown instead. Thus the Hoopoe was able to survive. In the Balochi language the Hoopoe is called Murgh-i-Suleman or King Solomon's 'bird'.

Food - Mainly insects, especially caterpillars.
Call - A very resonant *ho-po-po* which has given it its name–Hoopoe.

Coppersmith Barbet

Jungle Babbler

Brown-Headed Barbet

Jungle Babbler

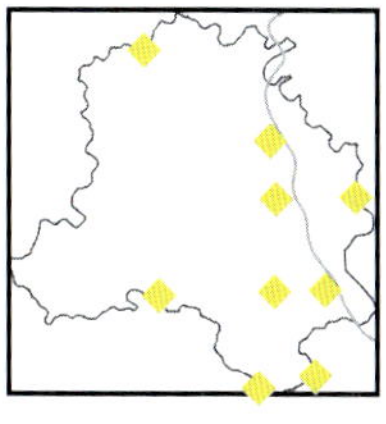

Size - *25 cm*

Hindi name - *Sat-bhai*

Scientific name - *Turdoides striatus*

Jungle Babblers are cross-looking birds, rather like some school teachers who think that the 'crosser' they look, the more easily they will be able to discipline their wards! These birds mostly remain in groups of six to twelve, which is why they are called Sat-Bhai or seven brothers in Hindi/Urdu.

The Jungle Babbler is a brown, messy and 'scruffy' looking bird. The upperparts are a slightly darker colour. The head is grey, and the bill is yellow. It has short rounded wings and an uncertain flight. Both sexes are similar in appearance.

The Jungle Babbler is a common bird of Delhi's gardens, orchards, jungles and parks.

The Pied-Crested Cuckoo lays its eggs in the Babbler's nest. The poor babbler raises the cuckoo chicks and can't make out the difference! That's another 'sly koel' for you.

Amusingly; Jungle Babblers are referred to as seven sisters in English! A tricky bit for our translators!

Food - Mainly insects; ants, cockroaches, moths, etc. It also likes fruit nectar.
Call - A loud and unpleasant *kae-kae kae* chattering sound.

Brown-Headed Barbet

Coppersmith Barbet

Common Tailorbird

Coppersmith Barbet

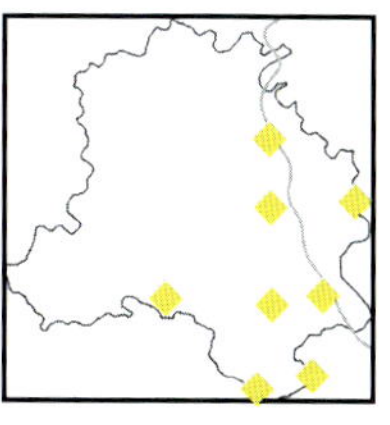

Resident Bird

Size - *17 cm*

Hindi name - *Chhota Basanta*

Scientific name - *Megalaima haemacephala*

Two types of Barbets are found in Delhi, the Coppersmith Barbet and the Brown-headed Barbet. The Coppersmith is the commoner of the two.

The Coppersmith Barbet is a big-billed bright green bird. It has a crimson coloured forehead and breast and a yellow throat. Both the sexes look alike, the juveniles though lack the crimson patch.

It is spotted mostly alone or in scattered groups on peepul trees. It is largely an arboreal bird and rarely comes on the ground.

The nesting season of the Coppersmith is from January to June. It nests in a tree hole, and lays about 2 to 4 dull white eggs. Both the sexes help in nurturing the young ones.

The call of the coppersmith barbet is similiar to the sound a coppersmith makes, hammering away at its utensils, hence the name.

Food - Mostly fruits and berries and occasionally insects like moths

Call - Its call is a repetitive *tuk... tuk... tuk* which is one of the most known bird calls in India.

COMMON TAILORBIRD

BROWN-HEADED BARBET

LESSER GOLDEN-BACKED WOODPECKER

Brown-Headed Barbet

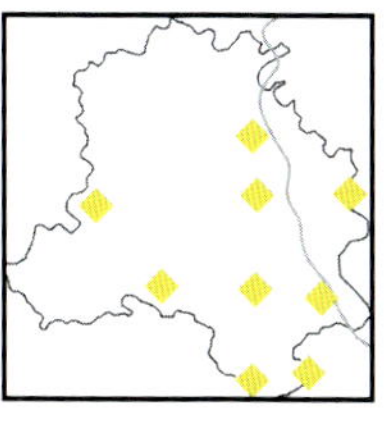

Size - *27 cm*

Hindi name - *Bada basanta*

Scientific name - *Megalaima zeylanica*

The Brown-headed Barbet is a plumpish bird. It has a strong and heavy bill. The overall plumage is green, the head, neck, upper back and breast are brown. There is a bare orange patch around the eye. Very 'earth' colours indeed! The sexes are alike in appearance.

The Brown-headed Barbet is a largely arboreal bird and can be spotted wherever there are big fruit trees. The Northern Ridge is a good place in Delhi. It can be difficult to spot because of the dense foliage, but its loud call can be heard from quite a distance. It is usually seen alone though it also collects in groups to feed together, sometimes in the company of the more common Coppersmith Barbet.

The nesting season of the Brown-headed Barbet is from February to June. Its nest is a hole excavated by the bird in a rotten branch of a big tree. The female lays 2 to 4 eggs. Both the sexes share the domestic duties.

Food - Ficus figs, berries, nectar, insects
Call - A monotonous *korr...r...r...kutrooo*

Lesser Golden-Backed Woodpecker

Common Tailorbird

Paradise Flycatcher

Common Tailorbird

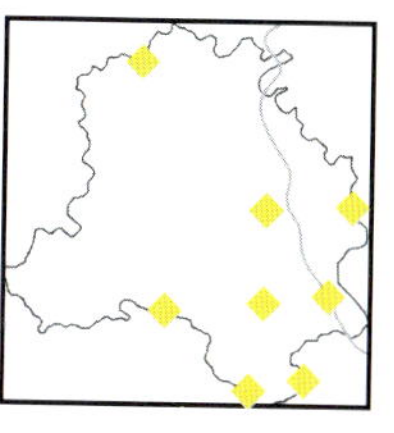

Resident Bird

Size - *13 cm*

Hindi name - *Darzi*

Scientific name - *Orthotomus sutorius*

Whenever I see the Common Tailorbird flitting around restlessly, I wonder if this tiny creature ever gets tired! It is an easily sighted bird in our gardens, parks and orchards and is a common resident of Delhi. The Tailorbird is at times seen singly, but more often in pairs.

The Tailor Bird is a small and jumpy dark green bird with a white belly and a rust-coloured crown. It is named so because it sews its nest by stitching together two or three leaves. The stitching is done with cobwebs, cocoon silk, wool or cotton. It uses its sharp beak like a needle to make holes in the leaves and uses strands of cobweb as a string. It even ties a knot at the end of the thread. That's one amazing tailor, who doesn't stitch clothes for others, but stitches a house for itself. Wouldn't it be more appropriate to call it the Masonbird! The nest appears like a deep cup. The nest is so well made that neither the rains nor the sun can damage it or the chicks inside it. The nesting season of the Common Tailorbird begins in June and continues throughout the monsoon upto August.

Mubashir Husain quotes Whistler as saying, "that like many other famous persons, the Tailor Bird is unsignificant in appearance."

Food - The tailorbird's favoured foods are small insects and their larvae
Call - Its call is a loud and rapid *pit-pit-pit.*

PARADISE FLYCATCHER

LESSER GOLDEN-BACKEI WOODPECKER

RED-BREASTED FLYCATCHER

Lesser Golden-Backed Woodpecker

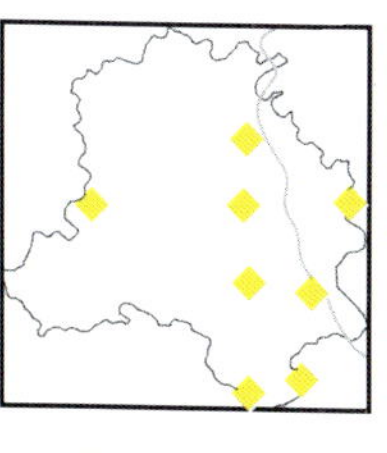

Resident Bird

Size - *29 cm*

Hindi name - *Katphora*

Scientific name - *Dinopium benghalense*

Of all the Woodpeckers, the Lesser Golden-backed Woodpecker is the most widely found in Delhi. It has a golden yellow back, with dull wings. The rump and tail are black. The underparts are white with dark markings. The head is white with a black nape and throat, and there is a grey eye patch. The adult male has a red crown. The female has a dark forecrown, with red only on the rear.

The Lesser Golden-backed Woodpecker is spotted alone or in pairs in open wooded forest, orchards and cultivation. Although it prefers large gardens and trees in big parks, many a time I have seen this 'carpenter' 'tuk-tukking' away at some tree in Buddha Jayanti Park.

The nesting season of the Lesser Golden-backed Woodpecker begins chiefly from March and goes upto August. Its nest is a hollow in a tree branch or a stem. It lays 3 glossy, milky-white eggs. Both the sexes share the domestic duties.

In English this bird only 'pecks' at the wood (Woodpecker), but in Hindi it 'breaks it apart'! Katphora means to break apart wood.

Food - Beetles, black ants, pulp of ripe fruits and flower nectar
Call - A chattering laughing sound

Red-Breasted Flycatcher

Paradise Flycatcher

Small Minivet

Paradise Flycatcher

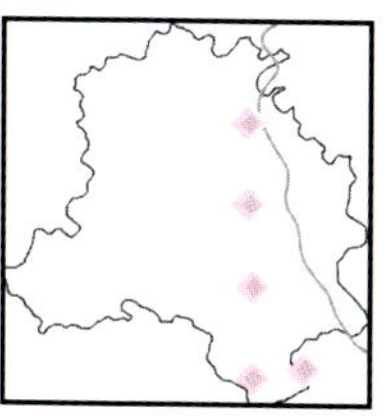

Summer Migratory Bird

Size - *14 cm*

Hindi name - *Shah bulbul*

Scientific name - *Terpsiphone paradisi*

The Paradise Flycatcher is a stunning-looking, silvery-white bird with a black crest on its head and a standout tail with two narrow beautiful feathers. The female does not have these streamers in the tail. The Paradise Flycatcher quite resembles the Bulbul in appearance.

The bird looks absolutely gorgeous while in flight. The delicate tail trails and loops around as it turns and hustles while catching insects.

The Paradise Flycatcher is found in shady groves, light wooded areas as well as around human habitation.

The nesting season of the Paradise Flycatcher is from February to July. The nest is a neatly woven cup of grass and fibres plastered outside with cobwebs. It lays 3 to 5, pale creamish-pink eggs, speckled with reddish-brown. Both sexes share the parental duties.

Why would we need a Flycatcher in Paradise? There wouldn't be any flies or other pests in Paradise and if there are they wouldn't be pests but little angels!

Food - Flies, gnats and other flying insects
Call - A grating *che-chew*, other sounds in the breeding season.

Small Minivet

Red-Breasted Flycatcher

House Crow

Red-Breasted Flycatcher

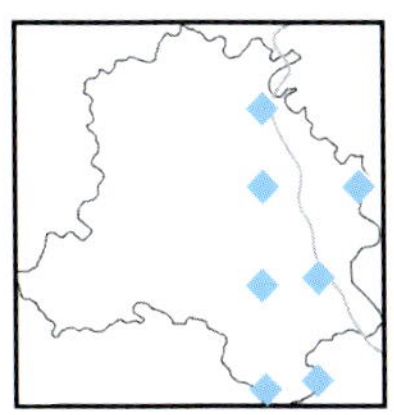

Winter Migratory Bird

Size - *13 cm*

Hindi name - *Turra*

Scientific name - *Ficedula parva*

The Red-breasted Flycatcher is a pretty little bird. The breeding male of the little flycatcher is mainly brown above and white below. It has a grey head and an orange breast. The bill is black and broad and has a pointed shape. Non-breeding males, females and fledglings have brown heads.

The Red-breasted Flycatcher is found singly in leafy trees, forests, plantations, parks, etc. It is a winter visitor to our region. For anyone spotting it for the first time, it will be difficult to identify immediately as it is a very restless bird and keeps flitting around. It also keeps cocking its tail up and down. That's a very comical sight indeed!

The Red-breasted Flycatcher is a migratory bird and does not nest in India.

Its breast is orange, wonder why it is called the Red-breasted Flycatcher!?

Food - Insects, caterpillars, berries, etc.
Call - A sharp click-click sound

House Crow

Small Minivet

Greater Coucal

Small Minivet

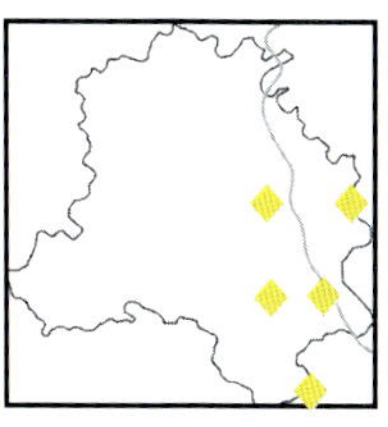
Resident Bird

Size - *15 cm*

Hindi name - *Rajalal*

Scientific name - *Pericrocotus cinnamomeus*

The Small Minivet has a strong dark beak and long wings. The male has grey upperparts and a grey head. It has yellowish-orange underparts, and orange tail edges, rump and wing patches. The female is grey above, with yellow underparts. Its tail edges, rump and wing patches too are yellow.

The Small Minivet forms small flocks. It is a widespread and common resident bird found in thorn jungle, scrublands, gardens groves, etc. It is a largely arboreal bird and rarely descends to the ground. So one has to first learn to recognise its call and then patiently scan trees and high shrubs and hope to spot it.

The nesting season of the Small Minivet is between February and September. Its nest is a neat little cup of fibres, etc plastered with cobwebs, high up in a tree. It lays 3 pale greenish-white eggs. The eggs may also be cream coloured, speckled with reddish brown. Both sexes share the domestic duties.

These scientific names really puzzle me. Why 'cinnamomeus'. It does not look like a piece of cinnamon nor is it found only on cinnamon trees! Then why?

Food - Insects and larvae
Call - A melodious and thin *swee swee swee*

Greater Coucal

House Crow

Koel

House Crow

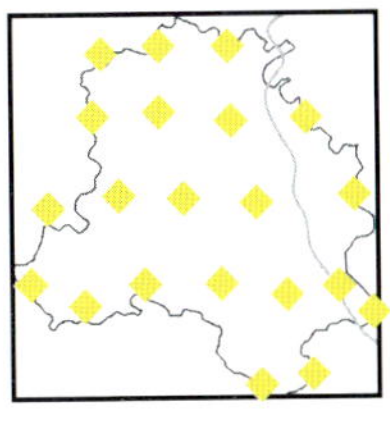

Size - *43 cm*

Hindi name - *Kowwa*

Scientific name - *Coruus splendens*

The House Crow is possibly the most common bird of India, and hence Delhi. It is found in nearly every area of the city. It is glossy black and grey. Both the sexes look alike.

The most valuable fact about the crow, from the ecological point of view, is that it is a scavenger. This means that it feeds on dead animals and so helps to keep the environment clean. It is a daring bird and steals food from houses and shops. A big bully, it is a nightmare for other birds and sometimes snatches food from bigger birds like Kites and Eagles.

The nesting season of the House Crow is mostly from April to June. Its nest is a platform of twigs. It lays generally 4 to 5 eggs. They are light bluish-green in colour, spotted with brown. Both the sexes share the domestic duties. The Koel regularly lays its eggs in crows' nests. It is probably the only bird which outfoxes the crow!

You must have read the story of the 'thirsty crow' which put pebbles into the pitcher to bring up the water. I tried it but could not get the water up. I wonder how many generations have been fooled by this story!

Food - Almost anything, from grains, seeds, fruits, to house scraps, leftovers and small animals like mice, frogs and lizards

Call - A loud *kown-kown*

Koel

Greater Coucal

Common Hawk Cuckoo

Greater Coucal

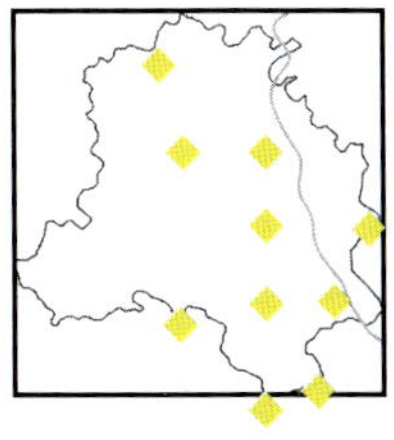

Size - *48 cm*

Hindi name - *Mahoka*

Scientific name - *Centropus sinensis*

The Greater Coucal is a big and bulky bird. It has a purplish-black head and body. The wings are brown above and black below, and the tail is dark green. This large bird's most striking feature is its eyes: blood-red and scary! Like the bloodshot eyes of a drunkard!

The Greater Coucal is seen alone or in pairs. It prefers to stay on the ground and is spotted in grasslands, open scrub forests and also near human habitation, leisurely strolling on the ground. You can stealthily follow it like a detective for a few metres, and it will go on walking ahead aware that you are trailing it. Finally, unable to shake you off, it will scuttle away into the undergrowth. In Delhi, you can see it in any of the larger and greener parks. The Northern Ridge is a good place to check it out.

The nesting season of the Greater Coucal is from February to September. It lays 3 or 4, dull white eggs. The nest is a large messy mass of twigs, leaves, etc. Both the sexes share the domestic duties.

Food - Insects like caterpillars, lizards, mice, bird eggs, etc
Call - A deep booming *coop...coop...coop*

COMMON HAWK CUCKOO

KOEL

PIED CRESTED CUCKOO

Koel

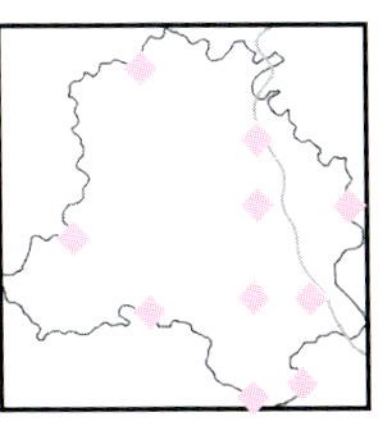

Summer Migratory Bird

Size - *43 cm*

Hindi name - *Koel*

Scientific name - *Eudynamys scolopacea*

The Asian Koel is a famous bird of India. In Hindi too it is called the Koel.

The size of a crow, the male Koel is a glossy black all over, with a yellow-green beak and ruby-red eyes. The female is dark-brown and white with spots all over with the exception of the belly which is grey. It is an easily spotted bird in Delhi's gardens and parks. Though it is more often heard than seen, I hear it in the district park very often on hot, sultry summer afternoons. The Koel is a locally migratory bird of India.

The nesting season of the Koel is from April to August. It is a parasitic bird and lays its eggs in the crow's nest. The eggs are smaller in size but similar to those of the crow. There is no fixed number of eggs laid. The male Koel distracts the attention of both the male and the female crows away from the nest. In the meantime the female Koel lays its eggs in the nest. The poor crows incubate and bring up the chicks, both theirs and the Koel's.

I never grudge the Koel its cheating because I sympathize with it for how Hindi/Urdu poets refer to it.The common refrain implies—though you are black (dark) you sing sweetly.What an obsession with 'fairness'.

Food - Ripe pipal, banyan and mulberry fruits, as well as insects, eggs of small birds, and nectar from flowers

Call - A noisy *koooo... koooo... koooo*, whistling call, which keeps on increasing in pitch and intensity.

PIED CRESTED CUCKOO

COMMON HAWK CUCKOO

BLACK DRONGO

Common Hawk Cuckoo

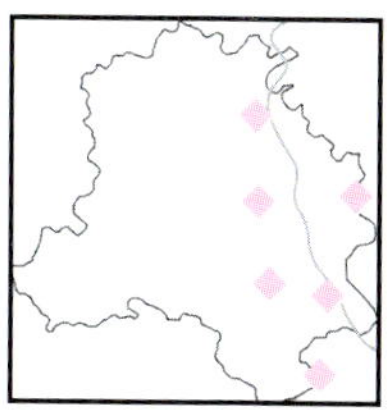

Summer Migratory Bird

Size - *34 cm*

Hindi name - *papeeha*

Scientific name - *Hierococcyx varius*

The Common Hawk Cuckoo is also known as the Brainfever Bird. It is a member of the Cuckoo family, and is parakeet-sized. It is smoky-grey, with white and brown upper parts.

The Common Hawk Cuckoo is spotted in groves, orchards and light wood forests near human habitation. It is a shy bird and is more often heard than seen. It can be heard all day and frequently on moonlit nights.

The nesting season of the Common Hawk Cuckoo is from March to June. As it is a parasitic bird, it does not make a nest but lays its eggs in the nests of birds like the Jungle and Common Babblers. It usually lays a single egg in each nest. The colour of the eggs is blue, similar to that of the Babbler. The hatching and rearing is done by the hoodwinked Babblers.

The call of the Common Hawk Cuckoo is said to bring rains, as it is more active during the summer season before the arrival of the monsoon.

Food - Caterpillars, berries, figs
Call - Loud scream; brain fever, brain fever… or in Hindi- *pee-kahan, pee-kahan...* Keeps rising in intensity.

BLACK DRONGO

PIED CRESTED CUCKOO

GREY FRANCOLIN (PARTRIDGE)

Pied Crested Cuckoo

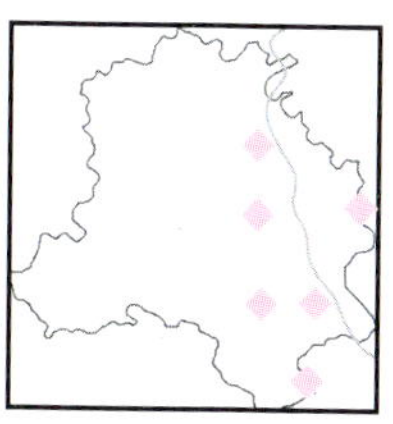

Summer Migratory Bird

Size - *33 cm*

Hindi name - *Kala Papeeha*

Scientific name - *Clamator jacobinus*

The Pied Crested Cuckoo is also known as the Jacobin Cuckoo. It is a beautiful bird, with a lovely crest on its head. It is black and white overall, with the tips of the tail feathers white.

The Pied Crested Cuckoo is usually spotted alone or in pairs. It is found in open, well-wooded areas, scrubs and large gardens. It is largely an arboreal bird but occasionally descends to the ground to hunt for grasshoppers and caterpillars.

The nesting season of the Pied Crested Cuckoo is from June to August. It is a parasitic bird, and like the Common Hawk Cuckoo lays its eggs in the nest of the Common and Jungle Babblers. It lays one egg, which is blue in colour, similar to that of the Babbler. The chick is brought up by the foster parents.

When Cuckoos don't make nests, how did we have a movie called 'One flew over the Cuckoo's Nest.'?

Food - Grasshoppers, caterpillars, berries
Call - A loud and metallic *piu..piu..pee…pee*

Grey Francolin
(Partridge)

Black Drongo

Eurasian Golden
Oriole

Black Drongo

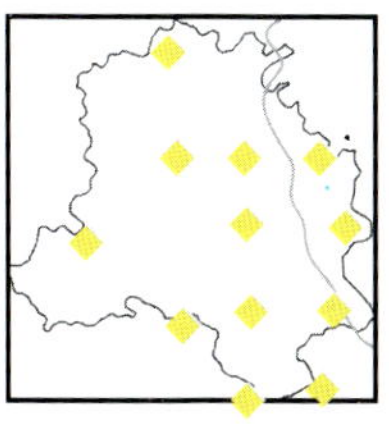

Size - *31 cm*

Hindi name - *Kotwal*

Scientific name - *Dicrurus macrocercus*

The Black Drongo is a slim, glistening black bird with shades of deep blue. It has a long forked tail, its distinguishing feature. It is in fact its tail which makes it one of the most easy birds to identify, even from a distance. It is an active bird. It can be spotted alone perched on telegraph wires and on grazing cattle keeping a lookout for insects. The Black Drongo is common throughout Delhi.

The nesting season of the Black Drongo is chiefly from April to August. Its nest is a fragile cup of slender twings held together with cobwebs. It lays 3-5 whitish eggs with brownish red stripes. Both parents aggresively guard the eggs and the fledglings.

The Black Drongo is a 'brave' and aggressive bird. In Hindi it is called Kotwal, meaning police inspector because it is said that Drongos sort out 'scuffles' between other birds. Also it protects the eggs of nearly every bird. Maybe, that is why birds like orioles nest near them! These gutsy forays have earned the drongo the name 'King Crow'.

Food - Almost exclusively insects, also flower nectar
Call - A harsh *tiu-tiu*

EURASIAN GOLDEN ORIOLE

GREY FRANCOLIN (PARTRIDGE)

INDIAN PEAFOWL

Grey Francolin (Partridge)

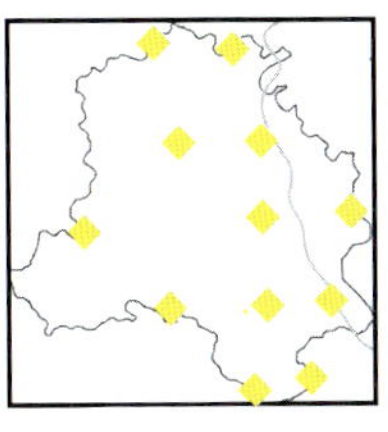

Resident Bird

Size - *33 cm*

Hindi name - *Safed yeetar*

Scientific name - *Francolinus pondicerianus*

The Grey Francolin was known as the Partridge until a few years ago, when it was rechristened. It has a brown back, with grey underparts, an orange face and a brownish patch on the belly.

It is spotted in pairs or groups in dry scrublands and near cultivation. It avoids dense and humid forests. The Asola Wildlife Sanctuary in Tughlaqabad, near Batra Hospital, is a good place to check it out. When disturbed, like most of the other partridges, it flies a little distance on rounded wings and follows up with a short glide.

Though hunting them now is illegal, Grey Francolins are still killed for food.

The nesting season of the Grey Francolin is probably all year round. The nest is a grass-lined scrape in scrubland, ploughed fields or grasslands. The Grey Francolin lays 4 to 8 eggs, which are creamish in colour.

Shooting down Partridges and Quails as well as ducks was a favourite sport with our erstwhile maharajas and nawabs. Collectively these birds were known as game birds.

Food - Grain, seeds, termites, beetle larvae, etc
Call - A loud and musical *pateela... pateela... pateela*

INDIAN PEAFOWL

EURASIAN GOLDEN ORIOLE

SPOTTED OWLET

Eurasian Golden Oriole

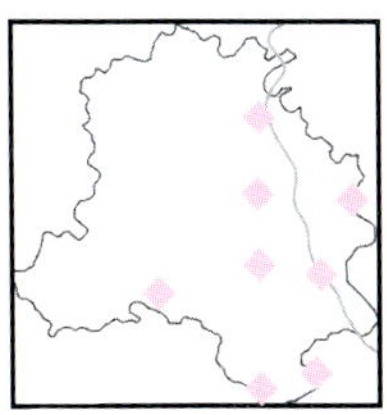

Summer Migratory Bird

Size - *25 cm*

Hindi name - *Peelak*

Scientific name - *Oriolus oriolus*

The Eurasian Golden Oriole is a gorgeously beautiful bird. The male has a striking persona with black and yellow plumage, and a black streak through the eye. The female though, is a dull green.

The Golden Oriole can be seen alone or in pairs. However, it is a shy bird and difficult to spot. The Golden Oriole is a largely arboreal bird and prefers open wooded country and groves around cultivation. In Delhi the best place to spot it is the Lodi Gardens. In fact this green lung of South Delhi is famous for the Golden Oriole.

The nesting season of the Golden Oriole is from April to July. The nest is a neatly woven cup made of grass bound with cobweb, suspended from a leafy twig. The Golden Oriole lays 2 to 3 eggs which are white, with reddish-brown spots.

A charming tale about the Golden Oriole tells us about a grove of mango trees which were always laden with ripe golden mangoes. People used to take as many as they needed. Once a greedy man started plucking too many. So the mangoes turned into golden birds and flew away! That was the birth of the Golden Oriole.

Food - Insects, flower nectar, peepul and banyan figs and other fruits and berries.
Call - A screech, but the song is a beautiful *weela-wee-ooo.*

Spotted Owlet

Indian Peafowl

Shikra

Indian Peafowl

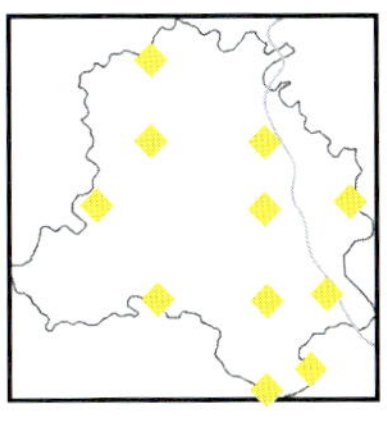
Resident Bird

Size - *Male- 92-122 cm, Female- 86 cm*

Hindi name - *Mor, Mayur*

Scientific name - *Pavo cristatus*

The Indian Peafowl is our very own National Bird, commonly known by the male of the species, the Peacock. It has a shining blue neck and breast, a wire-like crest and a lovely luxurious tail. The female lacks the shining blue of the male and is browner and duller. It also does not have the grand and gorgeous tail.

The Peafowl is a commonly sighted bird in green Delhi. The Northern Ridge, the Buddha Jayanti Park, the Deer Park, the JNU Campus and possibly other large green parks, are good places to spot this beautiful bird. It prefers areas near human habitation and near water. It is mostly spotted solitary and sometimes in groups. The Indian Peafowl is shy and can be tamed, but is also very cautious and wary of strangers. It is one of the first creatures to sense imminent danger.

The nesting season of the Peafowl is from January to October. It is during this season when the male spreads out its glorious feathers and struts around to attract the female. It nests on the ground in a shallow scrape lined with leaves and sticks. The female lays 3 to 5 shiny creamish-coloured eggs. The fledglings, rarely spotted, are very scrawny and dull looking unlike their magnificent papa.

The Indian tradition says that the Peacock calls 'piya...piya...' which means beloved for its mate. Its call conveys its yearning.

Food - Grain, vegetable shoots, insects, lizards, snakes etc.
Call - A very familiar and pleasing *mayon... mayon....*

Shikra

Spotted Owlet

Common Kestrel

Spotted Owlet

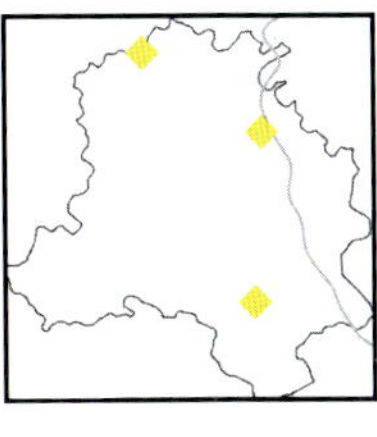

Size - *21 cm*

Hindi name - *Ullu*

Scientific name - *Athene brama*

The Spotted Owlet is a small and stoutly-built owl. The upperparts are grey-brown and spotted with white. The underparts are white and streaked with brown. The colour of the face is lighter than the body colour and the eyes are big and yellow. There is a pale white band around the neck.

The Spotted Owlet is a crepuscular bird but is sometimes seen during the day. It is a common resident bird of open habitats including cultivation and human habitation. It nests in holes in trees or crannies in ancient and big buildings. Look out for them in the Campus trees of Delhi University. Please look for them up in the trees and not below or behind them you might disturb other creatures!

The nesting season of the Spotted Owlet is from November to April. It makes it nest in old tree hollows or holes in dilapidated walls. It lays 3 or 4 white eggs. Both the female and the male share the domestic duties.

The scientific name of the Spotted Owlet, 'Athene Brama' is derived from the Greek goddess of wisdom 'Athena', after whom the city of Athens is named.

Food - Mainly insects like beetles, mice, lizards and young birds.
Call - A harsh *chirurr-chirurr-chirurr* Tawny Eagle.

Common Kestrel

Shikra

Oriental Honey-Buzzard

Shikra

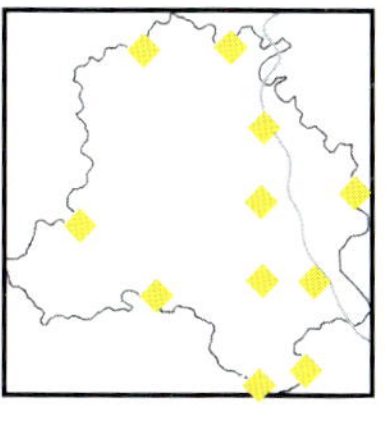

Resident Bird

Size - *30-34 cm*

Hindi name - *Shikra (male), Cheepak (female)*

Scientific name - *Accipiter badius*

The Shikra is a small but strong bird of prey. It has short broad wings and a long tail. The Shikra has pale grey upperparts, and is white and red below. The female is browner above. As in other hawks, the female is larger than the male.

The Shikra is a bird of open woodlands, including savannah and cultivation. It is a familiar sight in the city's parks and residential colonies. You are sure to see some at the Yamuna Bio-diversity Park, sitting calmly atop trees patiently watching for prey. The Shikra is spotted in pairs as well as alone. It is a swift flier, and usually flies close to the ground. Its flight has several quick wing strokes followed by a short glide.

The nesting season of the Shikra is chiefly from March to June. The nest is an untidy and loose collection of twigs, lined with roots and grass. It lays 3 to 4 bluish-white eggs. Both sexes share the domestic duties.

The word 'Shikra' is probably a derivative of the word 'Shikar', meaning to hunt.

Food - Lizards, dragonflies, squirrels, mice and small birds.
Call - A loud and harsh sound.

Oriental Honey-Buzzard

Common Kestrel

Booted Eagle

Common Kestrel

Winter Migratory Bird

Size - *36 cm*

Hindi name - *Karontia*

Scientific name - *Falco tinnunculus*

The Common Kestrel is smaller in size when compared to other birds of prey. Kestrels have long wings as well as a long tail. The Kestrel's plumage is brownish-grey with black spots. The male has a bluish-grey head and tail.

The Common Kestrel hunts during the day. It prefers fields and marshland and is mostly spotted single. When hunting, the Kestrel hovers, almost stationary, about 10-20 m above the ground, looking for its prey. Once it spots its prey, it makes a short dive towards the unfortunate creature and catches it with its razor-sharp talons.

The Common Kestrel's nesting season is from April to June, in the Himalayas. It does not nest in Delhi. Its nest is a mixture of twigs, roots and rags in a hole or a crevice of a cliff. It lays 3 to 6, pale pink or yellowish eggs, spotted with red. Both sexes share the parental duties.

Food - Small mammals, small birds, large insects, earthworms and frogs.
Call - A sharp *kii-kii-kii*.

BOOTED EAGLE

ORIENTAL HONEY-BUZZARD

PEREGRINE FALCON

Oriental Honey-Buzzard

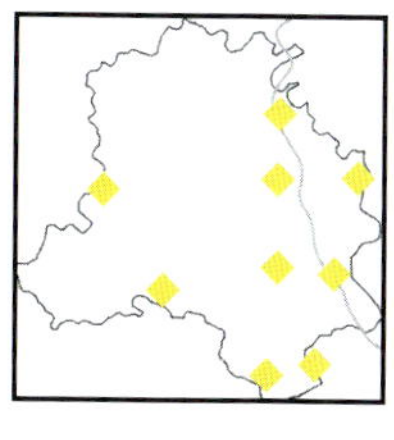
Resident Bird

Size - *68 cm*

Hindi name - *Shahutela*

Scientific name - *Pernis ptilorhyncus*

The Oriental Honey-buzzard is a majestic-looking bird. It has variable colouring. A greyish-brown with a darker head is the most common. It has silvery grey wings with dark barring at the tips and a greyish-rounded tail with broad, black terminal bands.

This buzzard is called the Oriental Honey-buzzard because it chiefly feeds on honey and bee larvae taken off live honeycombs. It occasionally eats small birds, frogs, insects and reptiles. It is normally spotted alone or in pairs.

This 'bear' of the avian world keeps a lookout for bees. When it spots some it follows them to their honeycombs. Then it digs up the 'nest' to feast on the honey, the main course accompanied by the larvae. The poor bees try in vain to sting it, but the thick feathers on its head act as a protective helmet.

The nesting season of the Oriental Honey-buzzard is between April and June. Its nest is a compact platform of sticks lined with leaves. It lays 2, dull cream or buff-coloured eggs, spotted with reddish brown. Both the sexes share the domestic duties.

Food - Honey, bee larvae, small birds, reptiles, frogs.
Call - A high-pitched screaming whistle.

Peregrine Falcon

Booted Eagle

Western Marsh Harrier

Booted Eagle

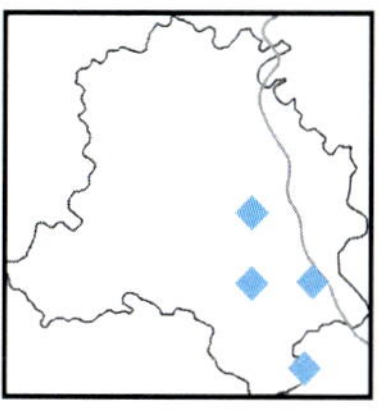

◆ Winter Migratory Bird

Size - *54 cm*

Hindi name - *Baghati*

Scientific name - *Hieraaetus pennatus*

The Booted Eagle is a handsome bird of prey. It is probably the smallest among all the eagles, and is rather kite-like in appearance. There are two plumage forms. The Pale birds are mainly light grey with a darker head and flight feathers. The other form has mid-brown plumage with dark grey flight feathers. It has feathered tarsus.

The Booted Eagle is a common winter visitor throughout India. It hunts in pairs, and roosts in groves of large leafy trees. This is a species of wooded, often hilly, country with some open areas. You can spot it at many places in Delhi, including the Ridge.

The Booted Eagle does not breed in our area, but in the Garhwal region. The nest is a platform of sticks high up in trees. Other nesting details are not clear.

All eagles have feathered tarsus, so I wonder why this particular specimen is called the Booted Eagle? In any case, the tarsus look more like thick woollen socks and not boots!

Food - Small mammals, reptiles and birds.
Call - A shrill *kli-kli-kli.*

Western Marsh Harrier

Peregrine Falcon

Indian White-Backed Vulture

Peregrine Falcon

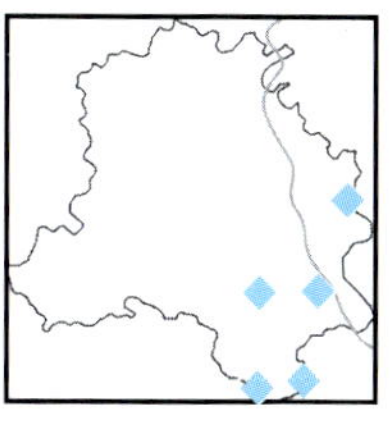

Winter Migratory Bird

Size - *38-48 cm*

Hindi name - *Bhyri(female), Bhyri bacha(male)*

Scientific name - *Falco peregrinus*

The Peregrine Falcon is a powerfully-built bird of prey. It has slaty, bluish-grey wings and a black back. It has a white face with a black stripe on each cheek and a blue-black head. It has large dark eyes, like a deer's, but nowhere as timid or a wee bit scared of any enemy! The female is larger than the male.

The Peregrine Falcon is usually spotted alone and likes to be around water. At the Yamuna, you can spot it terrorizing the ducks. It can also be spotted at historical monuments like the Qutab Minar, where it can be seen eyeing and then pouncing upon pigeons. Sadly, at the time of writing an individual is caged at the Delhi Zoo. That's a pretty insensitive thing to do.

The Peregrine Falcon is the fastest animal on the planet. Though its level flight is a bit faster than that of many other bigger birds, its diving speed is much greater. It has been measured reaching dive speeds of 350km/h! A true Olympian!

The Peregrine Falcon is a migratory bird and does not nest in India.

My first memory of a falcon is that of a very handsome one, named Allah Rakha, peiched on the aim of Amitabh Bachchan, in a movie.

Food - Ducks, partridges, pigeons.
Call - A loud scream.

INDIAN WHITE-BACKED VULTURE

WESTERN MARSH HARRIER

SMALL BEE-EATER

Mehran Zaidi

Western Marsh Harrier

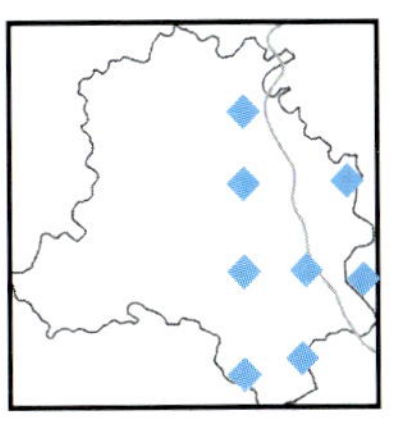

◆ Winter Migratory Bird

Size - *59 cm*

Hindi name - *Kutar*

Scientific name - *Circus aeruginosus*

The Western Marsh Harrier is an expert flier. The adult male has wings with grey and brown sections and black wingtips. Its head, tail and underparts are greyish, except for the chestnut-brown belly. The female is mainly brown with a cream crown and creamish edge to her wings.

Western Marsh Harriers are medium-sized raptors and are seen in marshes, jheels, paddy fields, etc. Marsh Harriers usually surprise their prey as they drift low over fields and reedbeds.

It's a fascinating experience to watch hundreds of ducks, lazily 'lolling' on a calm water body, suddenly take flight quacking and honking. You know that the Marsh Harrier has been sighted!

The Western Marsh Harrier does not breed in India.

Food - Small mammals, insects, birds, carrion, etc.
Call - None.

Small Bee-Eater

Indian White-Backed Vulture

Bay-Backed Shrike

Indian White-Backed Vulture

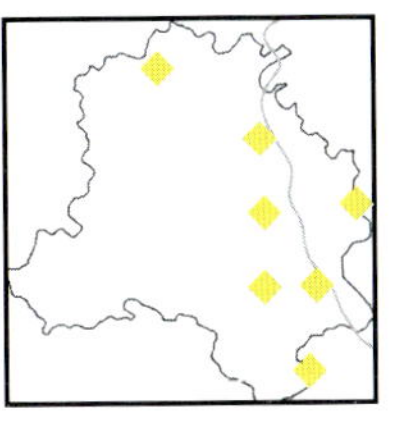

Size - *85 cm*

Hindi name - *Gidh*

Scientific name - *Gyps bengalensis*

Resident Bird

A majestic bird, the White-backed Vulture is a typical vulture, with a naked head, very broad wings and a short tail. It has a white neck. The adult's whitish back contrasts with the otherwise dark plumage.

The Indian White-backed Vulture is Delhi's commonest vulture. Like other vultures it is a scavenger, eating mostly from carcasses of dead animals, which it finds by soaring around human habitation. The White-backed Vulture is often seen in flocks, or rather was seen, as they have declined at a rapid pace. The most likely reason for their sudden disappearance is a disease, caused by a poisonous and toxic chemical, in the bodies of the carrion eaten by them. This chemical is diclofenac—a painkiller given to cattle. Vultures died so quickly that the drug was banned, but its effect is still felt.

The nesting season of the Indian White-backed Vulture is from October to March. Its nest is a large shabby platform of sticks on a banyan or some other big tree. It lays a single white egg, spotted with reddish brown.

It is believed that a flapping Vulture brings rain, while a soaring one shows coming heat and sun. That should be more dependable than our weathermen!

Food - Carrion.
Call - A loud screeching sound.

BAY-BACKED SHRIKE

SMALL BEE-EATER

COMMON WOODSHRIKE

Small Bee-Eater

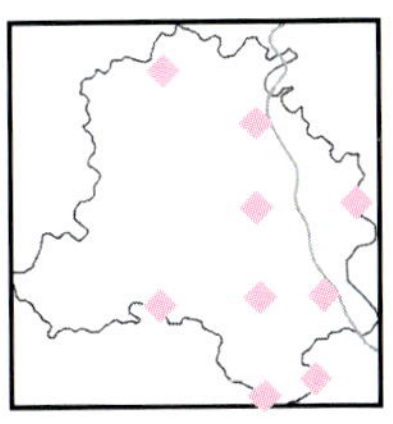

Summer Migratory Bird

Size - *21 cm*

Hindi name - *Patringa*

Scientific name - *Merops orientalis*

The Small Bee-Eater is a brightly-coloured, delicate and slim bird. It has green upper parts, whereas the head and the face are rufous or chestnut coloured. The underparts and the wings too are green. It has a black bill. The sexes are alike in appearance.

The Small Bee-Eater is a tame bird, abundantly found throughout the region. It is more commonly spotted in bushy lands, near cultivation, gardens, etc. It is also found close to water as it looks out for dragonflies. The Small Bee-Eater is often found in pairs but also sometimes forms small colonies, or nests near other Bee-Eaters.

The nesting season of the Small Bee-Eater is from February to May. Its nest is a horizontal tunnel dug into the side of an earth-cutting or into uneven sandy surfaces. It lays 4 to 7, milky-white eggs. Both sexes share the domestic duties.

What does the Bee eater say when it wonders whether it should have lunch or not? 'To be(e) or not to be(e) is the question!'

Food - Insects, especially bees, wasps, ants and dragonflies.
Call - A soft trill.

Common Woodshrike

Bay-Backed Shrike

Eurasian Wryneck

Bay-Backed Shrike

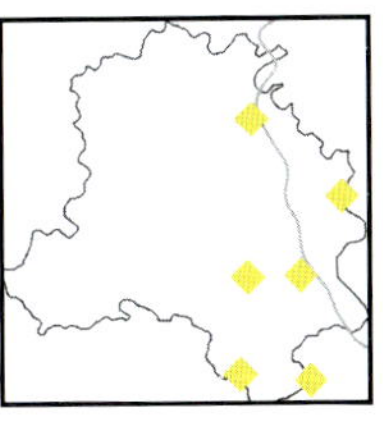

Resident Bird

Size - *18 cm*

Hindi name - *Pachanak*

Scientific name - *Lanius vittatus*

The Bay-backed Shrike is the smallest Indian Shrike. It is maroon-brown above with a pale rump and a long black tail edged with white. The underparts are white, but with buff flanks. The crown and the nape are grey, and the bill and legs are a darker grey.

The Bay-backed Shrike is found alone in scrubby areas, thinly-wooded country and cultivation. You can spot it sitting atop telegraph wires, looking out for its prey. The Sultanpur National Park, on the outskirts of Delhi, has reported good sightings of the Bay-backed Shrike.

The nesting season of the Bay-backed Shrike is from April to September. Its nest is a neat cup of grass, rags and feathers bound on the outside with cobwebs. It lays 3 to 6 eggs, which are a pale greenish-white, spotted with purplish-brown.

Food - Lizards, large insects, small birds and rodents.
Call - Harsh chirruping notes.

Eurasian Wryneck

Common Woodshrike

Bluethroat

Common Woodshrike

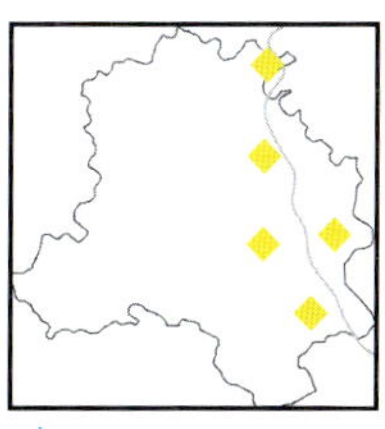
Resident Bird

Size - *18 cm*

Hindi name - *Keroula*

Scientific name - *Tephrodornis pondicerianus*

The Common Woodshrike is a dull, inconspicuous greyish-brown bird with a dark stripe below the eye. This bird is easily overlooked because of its dull plumage.

It is spotted in pairs or in groups in light scrub and bush forests as well as near cultivation. The Common Woodshrike is a familiar bird throughout the area.

The nesting season of the Common Woodshrike is between February and September. Its nest is a tidy-looking cup of soft bark and fibres cemented together and draped with cobwebs. It lays 3, pale green eggs, spotted with purple-brown. Both the sexes share the domestic duties.

Food - Insects, flower-nectar.
Call - A whistling and pleasant *whi… whi … whi.*

Bluethroat

Eurasian Wryneck

Brown Rockchat

Eurasian Wryneck

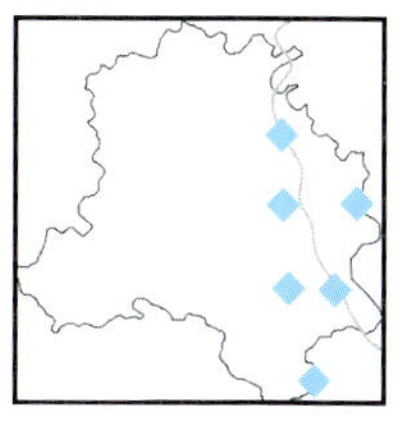

Winter Migratory Bird

Size - *19 cm*

Hindi name - *Gardan eyengtha*

Scientific name - *Jynx torquilla*

The Eurasian Wryneck is silver-brown above and whitish below. The sexes are alike in appearance.

The Eurasian Wryneck is a bird of open woodlands, thorny forests and orchards. It is mostly seen in sandy areas, where it forages for ants. It has an amusing way of stretching its neck and bill and twisting its head from side to side, hence its name. It does this mostly when it is startled or senses danger. The Eurasian Wryneck is mostly spotted solitary. In Delhi I have often spotted one near the temple on the banks of the Yamuna.

The Eurasian Wryneck does not breed in or rear Delhi but in Kashmir. Its nesting season is May and June. Its nest is a hollow on a rotten tree stem or branch. It lays 6 to 8 dull white eggs.

The Hindi name of this 'birdie' is a literal translation of its English name or maybe vice-versa, for it means 'neck twister'. Wrynecks, let me assure you, can never get cervical spondylitis, as they do so much of Brahma Mudra!

Food - Ants, etc.
Call - A *kek-kek-kek* sound.

Brown Rockchat

Bluethroat

Black Redstart

Bluethroat

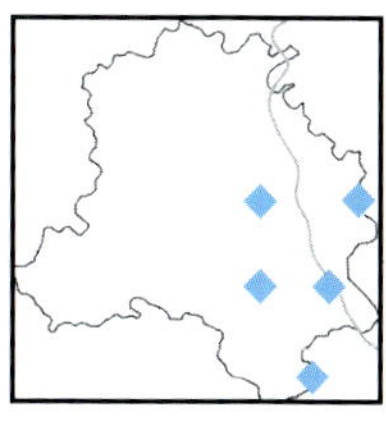

Winter Migratory Bird

Size - *15 cm*

Hindi name - *Hussaini pidda*

Scientific name - *Luscinia svecica*

The Bluethroat is a bubbly little bird. It is light brown, with a black tipped orangish tail. The adult male has a beautiful blue throat with a white patch in the middle. It looks absolutely exquisite, like a little artefact of Dresden china. The female has brown spots on a whitish breast.

The Bluethroat is found usually solitary in damp areas, water tank edges, sugarcane fields etc, hopping around with its cocked-up tail. If it gets used to humans, it can get bold and friendly. However, when startled or suspicious, it lowers its head and tail and scurries away, bobbing up its neck every now and then to look at the intruder. It's a very amusing sight!

The Bluethroat does not nest in the Delhi region, but in Ladakh. Its breeding season is normally June and July. Its nest is a cup of grass on damp ground, hidden amongst grass and scrub. The Bluethroat lays 4 dull green or dull brown eggs, spotted with pale red.

Food - Caterpillars, small beetles, other insects.
Call - A harsh *churr...chuk....*

Black Redstart

Brown Rockchat

Pied Bushchat

Brown Rockchat

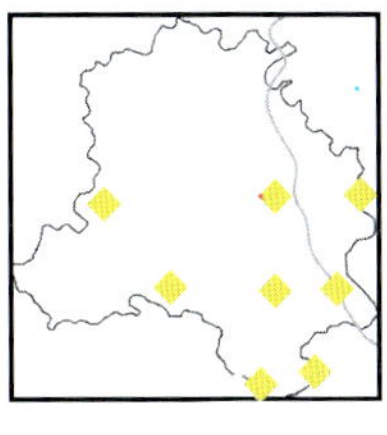

Size - *17 cm*

Hindi name - *Dauma*

Scientific name - *Cercomela fusca*

The Brown Rockchat is a plain-looking bird. It is dull-brown above, and rufous-brown below with darker wings and a blackish tail. The sexes are alike in appearance.

The Brown Rockchat is spotted alone or in pairs. It likes to be near rocky hills, ravines, etc. You can also spot it in your residential compound and even in your verandah. The Brown Rockchat is normally a tame and friendly bird, and likes to feed on the ground.

The nesting season of the Brown Rockchat is between February and August. Its nest is a cup made from roots, etc, placed in a rock cleft or in the hollows of walls. The female lays 3 to 4 dull blue eggs with orangish spots. It incubates alone.

Food - Insects.
Call - A short whistling *chee...chee...*

Pied Bushchat

Black Redstart

Lesser Whitethroat

Black Redstart

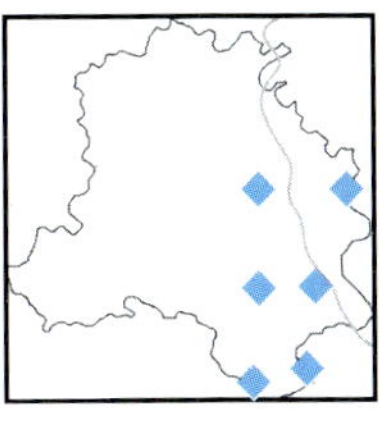

◆ Winter Migratory Bird

Size - *15 cm*

Hindi name - *Thirthira*

Scientific name - *Phoenicurus ochruros*

The Black Redstart is a small and energetic bird, and pretty as a picture. The male is black and orangish-brown in colour. The female is greyer and browner with a chestnut-coloured tail.

The Black Redstart is found alone in scarce scrubland and tree groves near cultivation, flitting around on trees constantly shaking its tail. Once a little fellow alighted on my Scorpio, when it was parked outside the Sultanpur National Park. I hope it had a good time, hopping all over it. I had for sure, 'watching' it and trying to film it with my Handycam.

The Black Redstart does not breed in the Delhi region but in the Himalayas. Its nest is a loose cup of grass, wool, feathers, etc. built under a rock. It lays 4 to 6, white or blue-green eggs.

'Thirthira', the Hindi name for the Redstart, means one who shivers!

Food - Insects, spiders, etc.
Call - A rattling song and a *tick-tick* call.

Lesser Whitethroat

Pied Bushchat

White-Throated Munia

Pied Bushchat

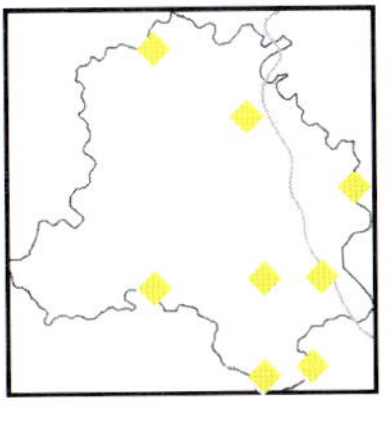

Size - *13 cm*

Hindi name - *Kala Pidda*

Scientific name - *Saxicola caprata*

The Pied Bushchat is a small and cute bird. The male is black except for a white rump, white wing patch and a whitish lower belly. The female has dark brown upperparts and chestnut brown underparts and rump. It does not have white wing patches.

The Pied Bushchat is a widespread and common breeding resident of the Delhi region. It is found in pairs in open habitats like scrublands, rough grassland, hillsides and cultivation near villages. You may also spot it in your neighbourhood. Just freeze and watch it. If you are in luck it might give you a treat, and sing for you.

The nesting season of the Pied Bushchat is from February to May. Its nest is a pad of grass lined with wool, in a hole in the ground or in an earth-cutting. It lays 3 to 5, pale bluish-white eggs, spotted with reddish-brown. The female incubates alone, but the male helps in building the nest and feeding the fledglings.

Food - Insects.
Call - A sharp *chek-chek* ending in a musical tweeet.

White-Throated Munia

Lesser Whitethroat

Wire-Tailed Swallow

Lesser Whitethroat

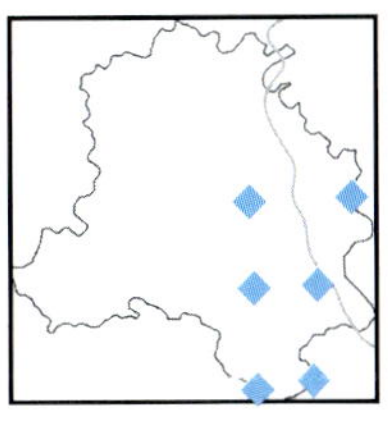

◆ Winter Migratory Bird

Size - *12 cm*

Hindi name - *not recorded*

Scientific name - *Sylvia curruca*

The Lesser Whitethroat is even smaller than the Pied Bushchat. It has a grey back, white underparts and a grey head. It has a coloured band across the eyes and of course a white throat! This tiny avian looks like a delicate little toy. The sexes are alike in appearance.

The Lesser Whitethroat is found alone in thorny scrub and babool trees near cultivation. In the Delhi region, you can find many of them at the Asola Wildlife Sanctuary in Tughlaqabad. The Lesser Whitethroat flies restlessly among tree branches and foliage, in search of insects.

The Lesser Whitethroat does not breed in Delhi.

Food - Insects, berries and other soft fruits.
Call - A low *tek-tek*.

Wire-Tailed Swallow

White-Throated Munia

Common Indian Nightjar

White-Throated Munia

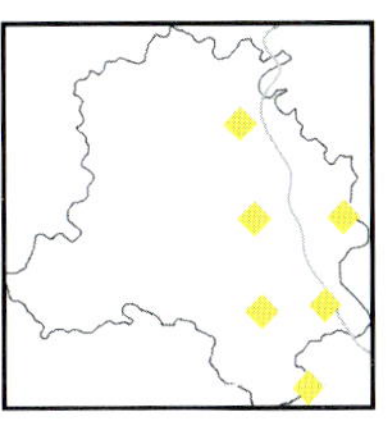

Resident Bird

Size - *10 cm*

Hindi name - *Pidda*

Scientific name - *Lonchura malabarica*

The White-throated Munia is also called the Indian Silverbill. It has a short silver-grey bill and brown upperparts. It has white underparts, a white rump and dark wings. The White-throated Munia has a long black tail. The sexes are alike in appearance. It's a pretty birds out of a fairytale.

The White-throated Munia is a gregarious bird found in pairs or in flocks in dry open country and cultivation, especially near water. In Delhi you can spot some at the Asola Sanctuary in Tughlaqabad and in any of the city's big parks. It avoids humid tracts.

The nesting season of the White-throated Munia is practically all year. Its nest is a big round structure of hard grass, lined with softer grass. This clever little bird also coolly uses the old, abandoned nests of the Baya Weaver bird to lay eggs. It lays 4 to 6, pearly-white eggs. Both sexes share the domestic duties.

Unlike other birds who abandon their nests after the chicks have grown up, the White-throated Munia uses them as family rooms!

The Munia is my favourite bird because my little sister's nickname is Munia and, if I may say so, she is as cute as the avian Munia

Food - Mainly seeds.
Call - A *tee-e tee-e faint* sound.

Common Indian Nightjar

Wire-Tailed Swallow

Paddyfield Pipit

Wire-Tailed Swallow

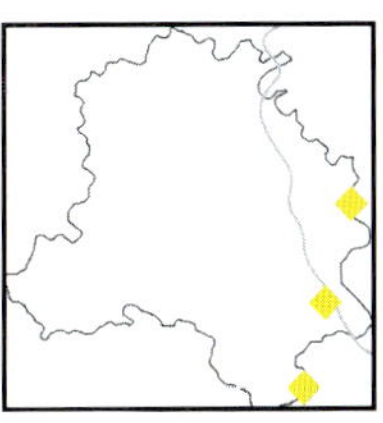

Size - *14 cm*

Hindi name - *Ababil*

Scientific name - *Hirundo smithii*

The Wire-tailed Swallow is a graceful and elegant bird. It has a glossy blue upper body, with a chestnut-coloured cap. The underparts are pearly white. The tail is unique. It has two 'wires' attached to it. The sexes are alike in appearance, except that the wires in the female are shorter. The wires are so thin that they may be difficult to spot from a distance.

The Wire-tailed Swallow is rarely seen in large flocks, but mostly in pairs or in small family groups. It roosts with other swallows in reedbeds and bushes. It is found in open areas and near water, mostly near tanks, streams, etc, where it hunts flying insects.

The nesting season of the Wire-tailed Swallow is usually all year, but it prefers the period between March and September. The nest is a saucer-shaped mud structure attached under a culvert. It also nests in closed verandahs. The female lays 3 to 5 white eggs, spotted with a deep reddish-brown. Both the sexes share the parental duties.

If you want to be safe from storms, from fires and do not want to be struck by lightening, invite a Swallow to nest on your roof, or so the legend goes!

Food - Aerial insects.
Call - A twittering *chik...chik...*

Paddyfield Pipit

Common Indian Nightjar

Large Pied Wagtail

Common Indian Nightjar

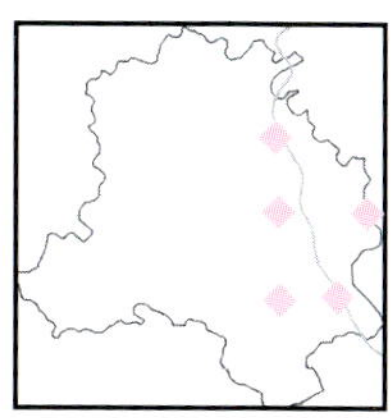

Summer Migratory Bird

Size - *24 cm*

Hindi name - *Chapka*

Scientific name - *Caprimulgus asiaticus*

The Indian Nightjar is a spooky-looking bird. It has buff and brown downy plumage which is soft. It has a white patch on each wing, Its wings are quite long.

Open woodland, scrub, and cultivation are the habitats of this nocturnal bird. The Indian Nightjar flies after sundown with an easy, calm flight. During the day, this Nightjar lies mute upon the ground. Due to its plumage it is difficult to spot, blending in with the soil. The Common Indian Nightjar frequently rests on roads during the night. This makes it easy to spot as it is easily seen by vehicle headlights.

The nesting season of the Indian Nightjar is not defined. It is anytime between February and September. It does not make a nest. It lays 2 pale pink eggs, spotted with reddish-brown. The eggs are laid on bare ground in a bush or a bamboo jungle.

Food - Moths, beetles and other insects.
Call - A familiar *chuk…chuk…chuk..r..r* sound.

Large Pied Wagtail

Paddyfield Pipit

White Wagtail

Paddyfield Pipit

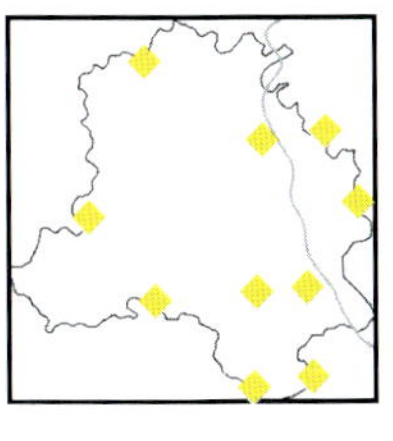

Resident Bird

Size - *15 cm*

Hindi name - *Rugail*

Scientific name - *Anthus rufulus*

The Paddyfield Pipit is a large pipit. It is otherwise a 'plain Jane', mainly streaked grey-brown above and pale below with streaks on the breast. It has long legs, a long tail and a long dark bill.

The Paddyfield Pipit is found in pairs or in scattered groups in open habitats, especially short grassland, cultivation, grazing land and ploughed fields. It is mainly a terrestrial bird.

The nesting season of the Paddyfield Pipit is between February and October. Its nest is a shallow cup of grass and roots, built on the ground. It lays 3 to 4, yellowish or greyish-white eggs, spotted with brown. Both the sexes share the parental duties.

Food - Weevils and other small insects.
Call - Characteristic '*chip-chip-chip*'.

White Wagtail

Large Pied Wagtail

House Sparrow

Large Pied Wagtail

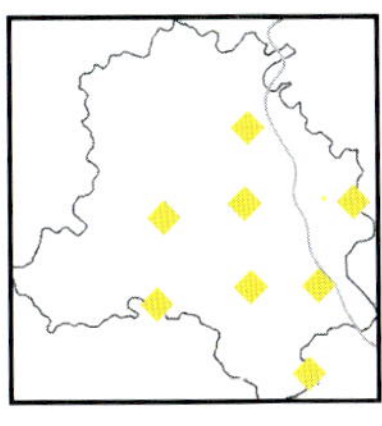

Resident Bird

Size - *21 cm*

Hindi name - *Mamula*

Scientific name - *Motacilla maderaspatensis*

The Large Pied Wagtail is also known as the White-browed Wagtail. It is an elegant-looking bird. It is black and white all over like the Magpie Robin, and has a prominent white eyebrow. The sexes are alike in appearance.

The Large Pied Wagtail is spotted around tanks, reservoirs, and house-tops. It continuously 'wags' its tail, like a happy dog! The Large Pied Wagtail is a conspicuous and noisy bird, although it is quite unafraid of humans. Its song too reminds me of the Magpie Robin.

The nesting season of the Large Pied Wagtail is not really fixed, but mostly it is anytime between March and September. Its nest is a pad of roots, wool, dry algae, hair etc, mostly under a bridge, always near water. It lays 3 or 4, greyish-brown or greenish-white eggs, speckled with brown. Both the sexes share the domestic duties.

In Manipur the Wagtail is considered an incarnation of Goddess Durga.

Food - Insects, spiders.
Call - Pleasant whistling sound.

Ashy Parinia

White Wagtail

Purple Sunbird

White Wagtail

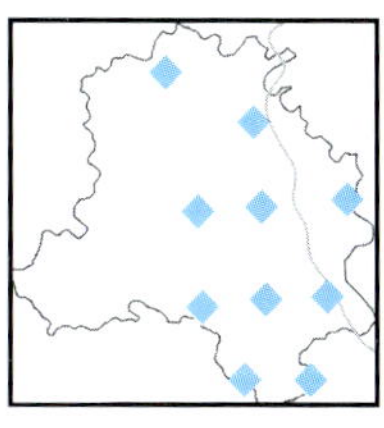
Winter Migratory Bird

Size - *18 cm*

Hindi name - *Peenchkkani*

Scientific name - *Motallica alba*

The White Wagtail is a good-looking bird about the size of the bulbul. It has a greyish back and rump and a white forehead. It has a black nape and a black breast. The belly is white. The sexes are alike in appearance.

The White Wagtail likes to be around water and damp areas, although it is less dependent on water as compared to other Wagtails. In Delhi, you can spot many flocks on the banks of the Yamuna, and also in big fields and golf courses. That's one 'high-society' bird, mingling with the page 3 crowd! Like its bigger cousin, the Large Pied Wagtail, the White Wagtail also continuously wags its tail.

The White Wagtail does not nest in our area. In India it nests only in Kashmir. The nesting season is from May to July. Its nest is a pad of roots, moss, hair, etc, in the broken walls of old buildings or amongst a pile of stones. It lays 4 to 6, white eggs, spotted with reddish brown.

For the ancient Greeks, the wagtail was symbolic of love. Maybe if socrates had carried a Wagtail he might have been spared.

Food - Insects, spiders.
Call - A sharp scream.

Water Birds

Darter

Great Cormorant

Great White Pelican

Cormorants

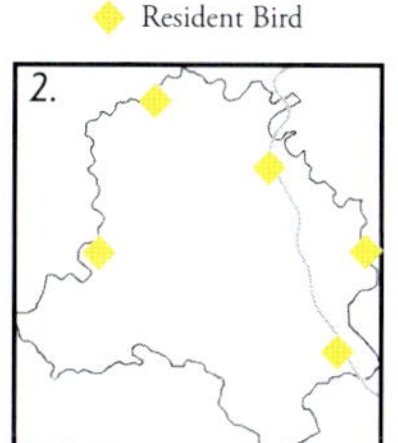

Resident Bird

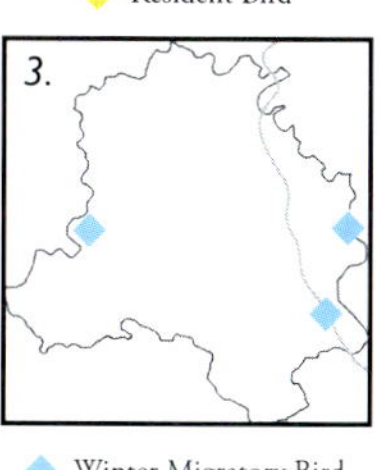

Resident Bird

3.

Winter Migratory Bird

Size:

Little Cormorant - 51 cm

Indian Cormorant - 63 cm

Great Cormorant - 80 cm

Hindi name - *Pan Kowwa*

Scientific names:

1. *Little Cormorant - Phalacrocorax niger*
2. *Indian Cormorant - Phalacrocorax fuscicollis*
3. *Great Cormorant - Phalacrocorax carbo*

Cormorants are expert fish catchers. The Little Cormorant is a glossy black water bird with a slim bill and a small white patch on the throat. The Indian Cormorant or the Indian Shag, as its also called, can be differentiated from the similar-looking Little Cormorant by its sleeker bill and lack of the shiny black plumage of the smaller bird. The Great Cormorant can be distinguished from other cormorants, during the non-breeding season, only by size. Breeding birds have a patch of white on the neck and the head.

The Indian Shag is more gregarious than the Little Cormorant and the Great Cormorant, and is often found in larger bodies of water and rivers. These Cormorants are mainly seen in jheels, rivers, etc. All the three Cormorants can be spotted at the Yamuna, the Yamuna Bio-diversity Park, and also at Sultanpur, usually in substantial numbers. The plumage of Cormorants is not waterproof.

The nesting season of the Little Cormorant is from July to September. Its nest is a shallow twig platform, usually in heronries. It lays 4 to 5, pale bluish-green eggs.

The Indian Shag's nesting season too is from July to September. Its nest is a shallow platform of twigs, located amongst mixed colonies of egrets, storks, etc. It lays 3 to 6, dull bluish-green eggs. Both sexes incubate.

The nesting season of the Great Cormorant is between September and February. The nest is a huge platform of twigs, often in mixed colonies. It lays 3 to 6, dull blue eggs. Both the sexes share the domestic duties.

Food - Almost exclusively fish.

Call - Usually silent except when close to the nest, it utters a variety of shrill sounds.

Great White Pelican

Darter

Purple Heron

Darter

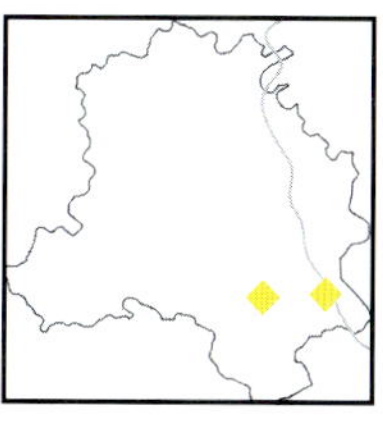
Resident Bird

Size - *90 cm*

Hindi name - *Panwa*

Scientific name - *Anhinga melanoguster*

The Darter is also called the snake bird because when it swims, only its head and its long, sleek neck are visible, and this makes it appear like a snake. It is mostly black in colour, with a white chin and a white throat. It has silverish-grey streaks on the back, a brown neck and a brown head. It is an expert juggler when it comes to feeding. It skews fish with its sharp beak and throws it up into the air. As the fish falls, it catches it and gulps it down whole!

The Darter is a familiar bird all over the Delhi region and can be spotted near tanks and jheels. Darters are not very sociable birds and prefer to be lone rangers. You will often spot a Darter sitting on a tree branch with it wings spread-out as if asking to be hugged! This is because like the cormorants and unlike other water birds its plumage is not impermeable. So, after getting wet, it has to sit on an appropriate perch, open out its wings and dry them up. Then only can it get back into the water and be able to swim.

The nesting season of the Darter is between June and August. Its nest is a twig platform similar to the nest of the Cormorant. It is built in trees near water and in mixed heronries. The Darter lays 3 or 4, dull greenish-blue eggs.

Food - Fish.
Call - Loud croaks and squeals.

Purple Heron

Great White Pelican

Indian Pond Heron

Great White Pelican

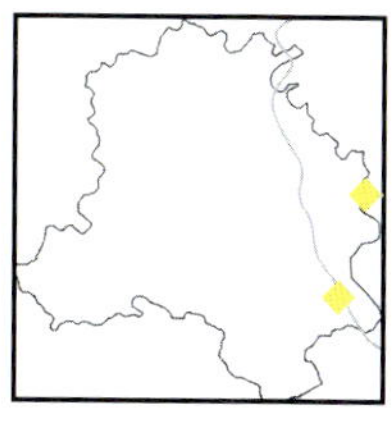

Resident Bird

Size - *183 cm*

Hindi name - *Hawasil*

Scientific name - *Pelecanus onocrotalus*

The Great White Pelican is also called the Rosy Pelican. It is a largely white bird, with shades of light red. It has a white crest, pink feet and a yellow clump on its breast. It has a brilliant yellow pouch on its lower bill. This pouch, which is like a fleshy bag is its most distinguishing feature. The skin around the eyes is pinkish.

It is totally a water bird which gathers in large numbers to feed. Sadly, in Delhi water pollution has turned these birds away from the Yamuna and other water bodies, where the sightings are now not very frequent. The Delhi Zoo is one place where you can find more than 40 individuals regularly. The Great White Pelican is a very fast and powerful flier.

The nesting season of the Great White Pelican is from February to April. The nest is a bed of feathers on the ground. It lays 2, pearly white eggs.

I am reminded of a cute limerick my mom once recited to me.

A wonderful bird is the pelican,
Its beak can hold more than its belly can.
It holds in its beak, enough food for a week,
I wonder how the hellican (hell he can)!

Food - Almost entirely fish.
Call - Rarely heard croaks and grumbles.

Indian Pond Heron

Purple Heron

Painted Stork

Purple Heron

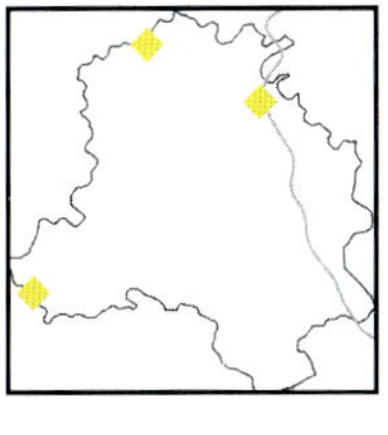

Size - *97 cm*

Hindi name - *Lal anjan*

Scientific name - *Ardea purpurea*

The Purple Heron is a handsome, wading bird. It has dark reddish-brown plumage, and, in the adult, a darker grey back. It has a sleek yellow bill, which is brighter in breeding adults.

The Purple Heron feeds in shallow water, catching its prey with its long, sharp bill. It will often wait motionless for its prey, or slowly stalk its victim. It tends to keep within reed beds more than the Grey Heron, and is often not visible, despite its size. It's probably shyer, or maybe more snobbish compared to its grey cousin!

Purple Herons are common throughout the Delhi region. You can spot them at the Yamuna, near the Okhla Barrage and at the Sultanpur National Park.

The nesting season of the Purple Heron is between June and March. Its nest is a platform of sticks in trees, reed beds, etc. It lays 3 to 5, pale, greenish-blue eggs. Both the sexes share the domestic duties.

Food - Fish, frogs, snakes etc.
Call - A loud and croaking sound.

Painted Stork

Indian Pond Heron

Asian Open-billed Stork

Indian Pond Heron

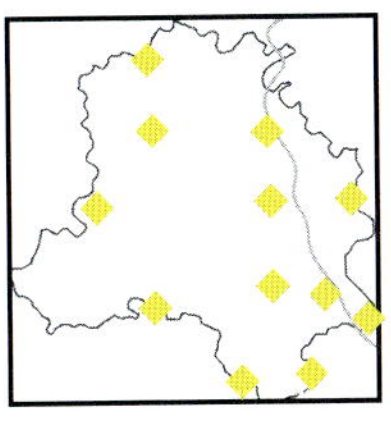

Resident Bird

Size - *46 cm*

Hindi name - *Bagla*

Scientific name - *Ardeola grayii*

The Indian Pond Heron is a small heron. This is a well-built species with a short neck, a short thick bill and a buff-brown back. In summer, the adults grow long neck feathers.

However, the Pond Heron looks different in flight. It appears very white because of the colour of its wings. The Indian Pond Heron's breeding habitat is marshy wetlands. This is a very common species in Delhi. It is a patient and an easily approachable bird. It can often be seen foraging around rubbish heaps.

The nesting season of the Pond Heron is between May and September. Its nest is an untidy twig platform. 3 to 5 eggs are laid which are pale greenish-blue in colour.

A white heron if sighted along with a black crow symbolises yin and yang of Chinese spirituality.

Food - insects, fish and frogs.
Call - A harsh loud croak uttered while flying.

Asian Open-billed Stork

Painted Stork

White Ibis

Painted Stork

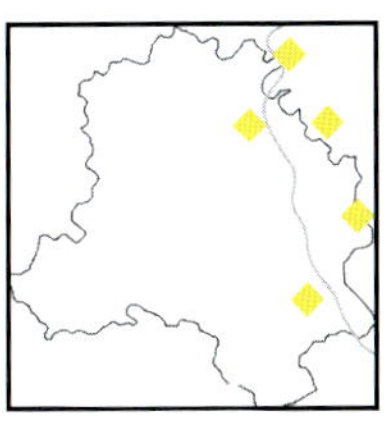
Resident Bird

Size - *93 cm*

Hindi name - *Janghil*

Scientific name - *Mycteria leucoephala*

The Painted Stork looks like the creation of an adventurous painter! It is white with greenish-black wings and tail quills, and a little pink on the shoulder and tail. A great yellowish-orange bill, red legs and a bare red face. Quite a kaleidoscope of colours!

The Painted Stork is a familiar bird found in lakes and near smaller water bodies of the NCR. It is normally spotted in small numbers. However, in a breeding colony, like the Delhi Zoo, you will see dozens of them creating a racket!

The nesting season of the Painted Stork is between August and January. The nest is a large platform of sticks lined in the middle with leaves, straw, etc. It is built on trees near water, or on small islands inside lakes. They breed in huge colonies called Heronries, often with other birds like Pelicans, Cormorants, Ibises, and other Storks. The Painted Stork lays 3 to 5, dull white eggs, sometimes with brown spots and streaks. The babies create a deafening din at feeding time.

Food - Mainly fish but also frogs, crabs and snakes.
Call - Mandible rattling, which is rattling of both the upper and the lower beaks together. Young in nests have raspy pleading sounds.

White Ibis

Asian Open-billed Stork

Black Ibis

Asian Open-billed Stork

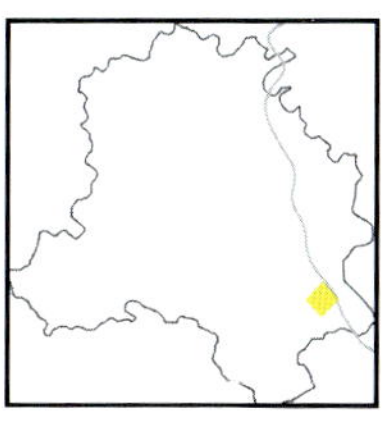

Resident Bird

Size - *81 cm*

Hindi name - *Gungla*

Scientific name - *Anastomus oscitans*

The Asian Open-billed stork is a bit smaller than its cousin the Painted Stork. It is a smooth white bird with black feathers and a reddish-black bill. The upper and lower parts of its bill (beak) don't close tight. A gap remains as you can see in the picture. That's why it is called the Open-billed Stork. This 'impolite' open-mouth structure however, allows it to eat snails.

The Asian Open-billed Stork is a very gregarious bird and nests in huge numbers. The Delhi Zoo is again a good place to spot it. It is a very powerful flier and flies at great heights. The Asian Open-billed Stork lives in marshes and jheels.

The nesting season of the Asian Open-billed Stork is between July and September. Like the Painted Stork it also breeds in heronries in the company of other water birds. The nest is a circular platform of twigs lined with leaves. It lays 2 to 4 white eggs.

In Greek mythology the stork is the 'mother', a nourisher.

Food - Frogs, Crabs and Snails.
Call - Mandible-rattling as in all storks.

BLACK IBIS

WHITE IBIS

EURASIAN SPOONBILL

White Ibis

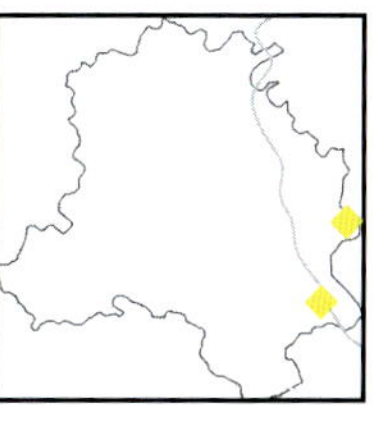

Size - *75 cm*

Hindi name - *Safed Baza*

Scientific name - *Threskiornis melanocephalus*

The White Ibis is a large white bird with a bare black, head and a black neck. Its beak too curves like its black cousin. It has black legs and a red patch under the wing.

The White Ibis is a gregarious bird and feeds and nests with other birds. In Delhi you can see it at the Delhi Zoo. It is very similar to the spoonbill in its habits. They are often found together.

The nesting season of the White Ibis is between June and August. Its nest is a platform of twigs in trees around water. The White Ibis lays 2 to 4 eggs, bluish or greenish white.

The White Ibis is similar to the Spoonbill in many ways, but it is also similar to the Peacock in one way. It is the national bird of an ancient country. Like the Peacock is the national bird of India, the White Ibis has the honour of being the national bird of Egypt.

Food - Frogs, insects, fish, snails and slugs.
Call - Like storks and spoonbills it does not produce any sound except nasal sounds heard during the breeding season.

Eurasian Spoonbill

Black Ibis

Greater Flamingo

Black Ibis

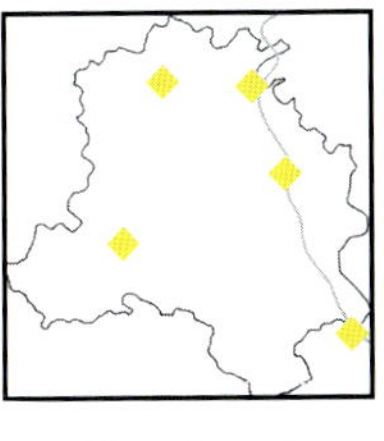

Size - *68 cm*

Hindi name - *Baza*

Scientific name - *Pseudibis papillosa*

The Black Ibis is of course a black bird with a long beak which curves downward. It has a white patch on the shoulder. It has red legs, and a red crown on its head.

The Black Ibis is found in marshes, jheels and also drier cultivation as it is not dependent on water like the White Ibis. In Delhi you might find some at the Yamuna. It feeds in shallow water alongside other ibises, spoonbills and storks. It is mainly spotted in pairs or in groups.

The nesting season of the Black Ibis is between March to October. Its nest is a large cup-shaped structure made from twigs and lined with straw, feathers, etc. The Black Ibis lays 2 to 4 eggs, which are bright, yet pale green, either unmarked or streaked with brown.

The Ibis represents the soul in Egyptian mythology.

Food - Insects, small reptiles and grains.
Call - two or three loud yells mostly while flying.

Greater Flamingo

Eurasian Spoonbill

Lesser Flamingo

Eurasian Spoonbill

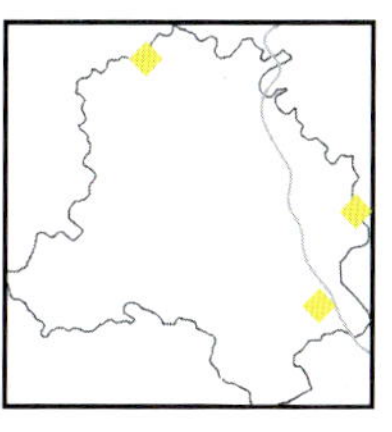

Resident Bird

Size - *60 cm*

Hindi name - *Chamach baza*

Scientific name - *Platalea leucorodia*

The Spoonbill is a snowy-white bird with long, black, legs and a large, flat spatula-type black bill with a yellow tip. It has a light brown patch on its neck. It develops a long crest during the breeding season. Makes it look pretty regal!

The flight of the Spoonbill is slow with steady wing beats. The Spoonbill is normally found near jheels, marshes and mud banks (both male and female), like the Yamuna and the Sultanpur Jheel.

The nesting season of the Eurasian Spoonbill is between July and November. Its nest is a big platform of sticks. Like the Stork, the Spoonbill too nests in heronries on trees near water. It lays 4 eggs which are white in colour and spotted with a deep reddish-brown.

Food - Tadpoles, frogs, snails, insects, plants and weeds.
Call - Clattering of mandibles and sometimes a low grunt.

Lesser Flamingo

Greater Flamingo

Greylag Goose

Greater Flamingo

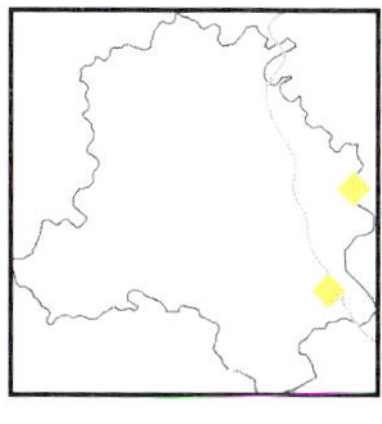

Resident Bird

Size - *140 cm*

Hindi name - *Bog hans*

Scientific name - *Phoenicopterus ruber*

The Greater Flamingo is a long-legged, long-necked bird. It is rosy pink with a large pink turned-down bill.

The Greater Flamingo is mostly spotted in parties or in small flocks on jheels, rivers, lagoons, mudflats, etc. Flamingos use their large beaks to sieve the small food particles from the water. A flamingo lowers its head into the water and moves it from side to side, collecting the food and water mixture. The fleshy tongue pushes the mixture past the tooth-like ridges on the outside of the beak and the lamellae, or the filters inside the beak. The lamellae separate the food particles from the water.

The nesting season of the Greater Flamingo depends on the shallowness of water on the nesting ground. The nest is a truncated conical mound of hard mud. The Greater Flamingo lays 1 or 2 white eggs with a slight bluish shade.

The flamingo's pink feathers are due to its diet, which is high in alpha and beta-carotene.

Food - Crabs, worms, insect larvae, marsh plant seeds, etc.
Call - A noisy goose-like honking sound.

Greylag Goose

Lesser Flamingo

Bar-Headed Goose

Lesser Flamingo

Winter Migratory Bird

Size - *105 cm*

Hindi name - *Chhota raj hans*

Scientific name - *Phoenicopterus minor*

The Lesser Flamingo's smaller size, darker-pink plumage and darker bill differentiates it from its Greater cousin. However, it is no way 'lesser' in being an impressive, handsome avian!

The Lesser Flamingo is a very gregarious bird and is often spotted in large flocks with the Greater Flamingo, in lakes, lagoons, etc. However, it has a preference for very saturated salt water areas, and that is why it is rarely spotted in the Delhi region. There have been a few sightings in the past, but not regularly. Its feeding style is similar to the Greater Flamingo.

The nesting season of the Lesser Flamingo is from July to January. It breeds in the Great Rann of Kutch like its Greater cousin, and its nest and eggs are similar to the Greater Flamingo.

Few sights in nature can parallel the beautiful vision of hundreds of flamingoes standing in water and the rising sun sparkling on the water, matching its glow with that of the birds

Food - Algae, diatoms, etc.
Call - Similar to the Greater Flamingo.

Bar-Headed Goose

Greylag Goose

Brahminy Shelduck

Greylag Goose

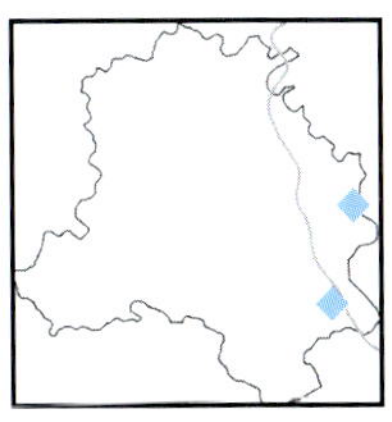

◆ Winter Migratory Bird

Size - *81 cm*

Hindi name - *Raj hans*

Scientific name - *Anser anser*

The Greylag is said to be the ancestor of all our domestic breeds. It is a happy fact that it did not 'lag' behind like the ancestors of other species; mammoths for instance. It is a large greyish-brown goose. It has a big head, pink legs and an almost triangular pink bill. Isn't that cute?

The Greylag Goose is a gregarious and alert bird spotted in marshes, jheels and lakes. A unique and interesting feature of the Greylag is that it rests for most of the day and feeds during the night. A very unhealthy habit, our dieticians would say! It is a winter visitor to the Capital region. About 500 to 600 of these birds can be easily sighted at the Yamuna and at the Sultanpur National Park.

The Greylag Goose breeds in Central Asia and not in India.

It foxes me why a goose should be called 'hans' in Hindi/Sanskrit, for hans means swan. By the way, 'Raj Hans' mean royal swan.

Food - Grass, gram and wheat shoots and aquatic tubers.
Call - A loud cackle *ahang-ung-ung*, like the domestic goose.

Brahminy Shelduck

Bar-Headed Goose

Cotton Teal

Bar-headed Goose

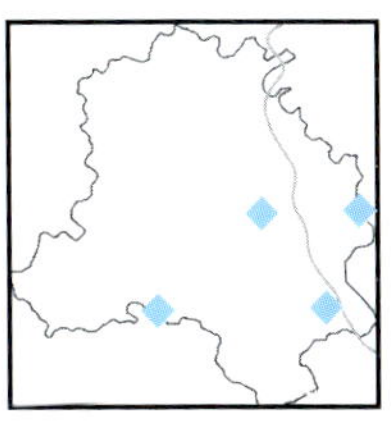

Size - *75 cm*

Hindi name - *Hans*

Scientific name - *Anser indicus*

The Bar-headed Goose is grey and white with two brownish-black bars on the rear of its white head. So now you know why its called the Bar-headed Goose, nothing intoxicating here! The body is grey by and large, and the bill and legs are pink, orange, or yellow.

The Bar-headed Goose is found in southeast Russia, Ladakh, Tibet and western China throughout the breeding season and in northern India and northern Myanmar during winter. The Bar-headed Goose sometimes rests along with Greylags in large flocks in the middle of the sandbanks of large jheels and rivers like the Yamuna. It is a very alert bird and hence difficult to outsmart.

The Bar-headed Goose does not nest in India, but in Tibet and Ladakh. Its nest is a depression in dense herbage on the edge of big lakes. It lays 3 to 4, pearly-white eggs.

The word 'goose' has somehow always conjured up the image of a very belligrent, aggressive creature for me. Probably because of Goosey Goosey Gander who threw an old man down the stairs because he wouldn't say his prayers!

Food - Soft gram and paddy shoots.
Call - A nasal musical sound.

Cotton Teal

Brahminy Shelduck

Gadwall

Brahminy Shelduck

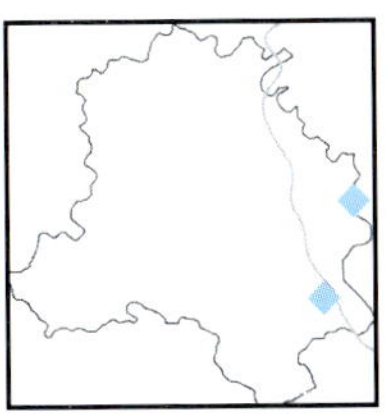

Winter Migratory Bird

Size - *66 cm*

Hindi name - *Surkhab*

Scientific name - *Tadorna ferruginea*

A dazzling orange and brown duck, the Brahminy Shelduck is a gorgeous specimen of the bird family. The male grows a black band around the neck in the breeding season.

The Brahminy Shelduck is a common winter migrant to large open lakes, river banks and reservoirs. Within India it breeds in high altitude lakes and marshes in Ladakh. It arrives in the Delhi region by October and leaves by April. I have spotted quite a few even in the small lake in Model Town, North Delhi and they were there for days. Lucky people of Model Town!

The nesting season of the Brahminy Shelduck is between April and June. Its nest is a solid pad of feathers in holes in cliffs. It lays 6 to 10, milky-white eggs.

The down feathers of this gorgeous duck are very soft. Hence over the years thousands have been killed to make super-soft pillows for the heads of rajahs and zamindars

Food - Water plants, insects, snails, fish, crabs and small reptiles.
Call - A nasal *aang, aang.*

Gadwall

Cotton Teal

Mallard

Cotton Teal

◆ Summer Migratory Bird

Size - *32 cm*

Hindi name - *Girri*

Scientific name - *Anas crecca*

The Cotton Teal is the smallest duck in the world. Chho chhweet! Its plumage is mostly white, with a blackish-green crown. Its short bill resembles that of a goose, which is why it is also called the Cotton pygmy-goose.

The Cotton Teal is not a very common bird in this area though the Yamuna is a good place to spot some. It can be seen in pairs or in small flocks on jheels, rivers, irrigation tanks, etc. It is a swift flier.

The nesting season of the Cotton Teal is between July and September. Its nest is a natural hollow in a tree trunk in or near water. It lays 6 to 12, pearly-white eggs.

Food - Water plant seeds, tender shoots, grains, water crabs, shrimps, worms, etc.
Call - An unusual *cluck-cluck* sound.

MALLARD

GADWALL

NORTHERN SHOVELER

Gadwall

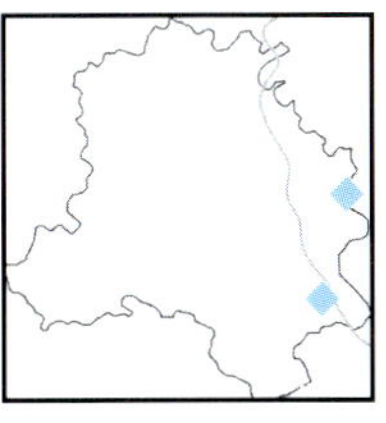

Winter Migratory Bird

Size - *51 cm*

Hindi name - *Myla*

Scientific name - *Anas strepera*

The Gadwall is a very sober-looking duck. The breeding male is a beautifully patterned grey, with a black back. The female is light brown, like the subdued but elegant Gadwal cotton sarees.

The Gadwall is not as gregarious as some other ducks, outside of the breeding season and tends to form only small flocks. It is a bird of open wetlands, lakes, wet grassland or marshes with dense vegetation.

You can spot large flocks at the Yamuna and the Sultanpur National Park. It usually feeds with its head underwater dabbling for plant food.

The Gadwall is a migratory bird and does not breed in India.

Food - Tubers and shoots of water plants, grains, seeds, etc.
Call - The male makes a harsh whistling sound, whereas the female has a typical duck like quack.

Northern Shoveler

Mallard

Spot-Billed Duck

Common Redshank

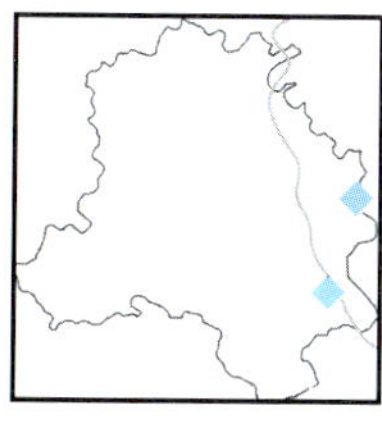

Winter Migratory Bird

Size - *28 cm*

Hindi name - *Chota Batan*

Scientific name - *Tringa totanus*

The Common Redshank is not very similiar to the Common Greenshank. The upper parts of its body are brownish and the rump is reddish. The inverted bill too is reddish. The legs are red, hence the name 'redshank'.

The Redshank is a wary but noisy bird which would alert others with a loud call, if it senses danger.

The Common Redshank breeds in the North-western Himalayas, but is a winter visitor to the North-western plains of India.

The nesting season of the Common Redshank is from May to July. It lays 4, yellow and reddish-brown eggs.

Food - Water insects, crabs, snails, slugs, etc.
Call - A loud and musical *tiwee-tiwee-tiwee.*

WHITE-BREASTED KINGFISHER

BLACK-WINGED STILT

WHISKERED TERN

Black-winged Stilt

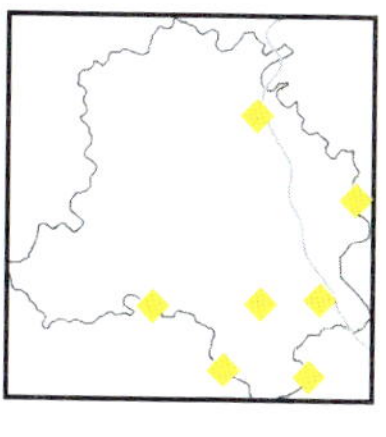

Size - *25 cm*

Hindi name - *Gaz paon*

Scientific name - *Himantopus himantopus*

The Black-winged Stilt is a common water bird of Delhi. It is a big black and white bird with long orangish-red legs and a black bill. It has black on the back of the neck, a white collar and red eyes.

The Black-winged Stilt swims well but is a weak flier. It is a social bird, and is usually found in small groups. Black-winged Stilts prefer freshwater and saltwater marshes and the shallow edges of lakes and rivers.

The nesting season of the Black-winged Stilt is between April and August. Its nest is a depression in the ground on the edge of a jheel or a raised platform of stones in low water, lined with vegetable froth. It lays 3 or 4 dull-coloured eggs, somewhat like those of the Red-wattled Lapwing.

The Hindi name of the Black-winged Stilt is very interesting. 'Gaz' is the obsolete Indian measure of length, about 90 cm, and 'paon' of course means legs. So, gaz paon means one with legs measuring a gaz!

Food - Worms, insects, snails, slugs, etc.
Call - A repeated sharp barking sound.

WHISKERED TERN

WHITE-BREASTED KINGFISHER

SARUS CRANE

White-breasted Kingfisher

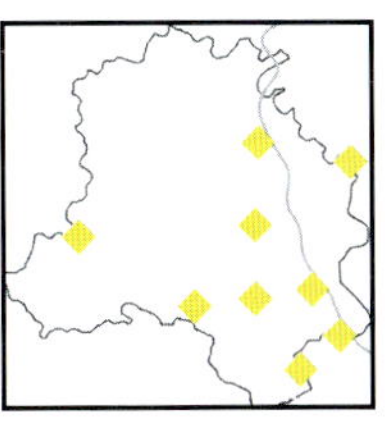

Resident Bird

Size - *28 cm*

Hindi name - *Kilkila*

Scientific name - *Halcyon smyrnensis*

The White-breasted Kingfisher is a brilliantly-coloured, beautiful bird. It is turquoise blue with a cocoa-brown head, neck and underparts. The breast is white. It has a heavy red bill.

This Kingfisher is spotted alone in both cultivated and wooded areas, near and away from water. It is the most common Kingfisher found in the Delhi area. It can be seen near ponds, big puddles, and rivers.

The nesting season of the White-breated Kingfisher is between March to July. It builds its nest in a horizontal tunnel dug into the side of a dry nullah. It lays 4 to 7, white, spherical eggs.

This bird should be extremely grateful to industrialist Vijay Mallya for immortalizing it, whether zipping across the skies or making men zip around after downing it!

Food - Fish, tadpoles, lizards, grasshoppers and other insects.
Call - A loud cackle.

Sarus Crane

Whiskered Tern

Grey Heron

Whiskered Tern

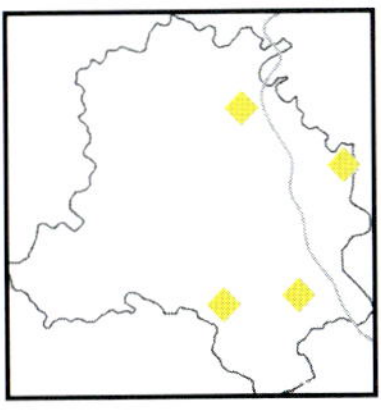
Resident Bird

Size - *25 cm*

Hindi name - *Tehari*

Scientific name - *Chlidonias hybridus*

The Whiskered Tern has a black cap, a strong bill and a short tail. It has dark grey breeding plumage. In summers it has white cheeks and red legs as well as a red bill. In winter, the forehead becomes white and the body plumage a dull grey. The sexes appear alike.

The Whiskered Tern has a very elegant flight, and an effortless glide. It can be spotted near jheels, rivers, marshes, etc.

The nesting season of the Whiskered Tern is between June and September. Its nest is a rough circular pad of reeds and rushes on floating water chestnut in jheels and swamps. It lays 2 to 3, dull yellow or greenish eggs speckled with purple and blackish-brown. Both the sexes share the domestic duties.

The Whiskered Tern has a very illustrious cousin—The Arctic Tern which holds the world record for flying the longest distance.When it migrates each year, it flies from the Arctic to the Antarctic and back again, back and forth, 40,000 km! In its lifetime it flies the same distance as to the Moon and back. Some reflected glory for the Whiskered Tern to bask in.

Food - Small fish, crabs, tadpoles and insects.
Call - A characteristic krekk.

GREY HERON

SARUS CRANE

CATTLE EGRETS

Sarus Crane

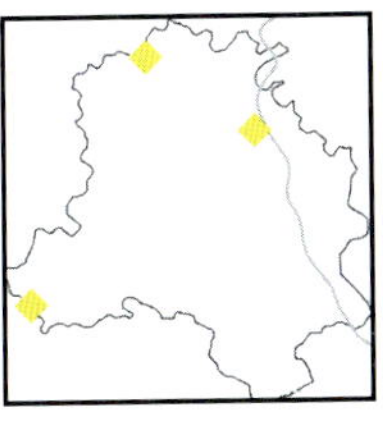

Resident Bird

Size - *156 cm*

Hindi name - *Sarus*

Scientific name - *Grus antigone*

The Sarus Crane is a very beautiful and graceful bird. It is light grey, with a bare red head, red legs and red upper neck. Grey and red has always been my favourite colour combination.

The Sarus Crane can be spotted in fields of crops. However, it has a preference for watered areas and marshes. In the Delhi region it can be spotted in the Sultanpur Sanctuary and also near the Najafgarh Canal. It is not totally a water bird and is often spotted away from water. The Sarus Crane is seen in pairs or in family groups. It feeds with other water birds. The Sarus Crane is a world champion, being the tallest flying bird in the world.

The nesting season of the Sarus Crane is between July and December. Its nest is a huge mass of reed stems and straw in the middle of a paddy field or a marsh. It lays 2 pale greenish or pinkish-white eggs. It is said that Sarus Cranes pair for life.

Once travelling from Delhi to Lucknow by the Shatabdi, the train groaned to a halt in the middle of nowhere. But it turned out to be a blessing in disguise as I was treated to the famous dance of a Sarus Crane pair in the fields just across.

Food - Fish, frogs, insects, crabs, grains and tubers.
Call - A very loud and far-reaching trumpeting sound.

CATTLE EGRETS

GREY HERON

COMMON SANDPIPER

Grey Heron

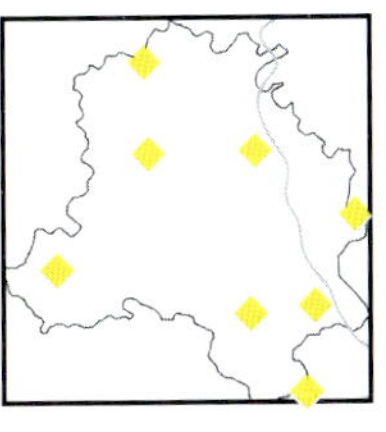

Size - *98 cm*

Hindi name - *Nari*

Scientific name - *Ardea cinerea*

The Grey Heron is a tall and long-necked bird. It is ashy-grey above, with a white crown and a black stripe between its eyes, like a dirty piece of band-aid on the nose bridge of a naughty boy!

The Grey Heron is by and large a lonely bird, which inhabits open marshes. It is a familiar figure near ponds and streams. Visit the Yamuna and you can spot many Grey Herons standing absolutely still, as if someone has yelled 'Statue!' at them! They generally stand 'frozen' waiting for their prey to come nearer. The Grey Heron flies with steady wing beats.

The nesting season of the Grey Heron is between July and September. Its nest, built in trees, mostly amongst mixed heronries, is a twig platform lined with grass, etc. The Grey Heron lays 3 to 6, greenish-blue eggs. Both the sexes share the domestic duties.

In English, men and women collectively, much to the chagrin of feminists, are called 'Man'. Well some small consolation; the Grey Heron male and female, is called 'Nari' (woman) in Hindi!

Food - Insects, fish and frogs.
Call - A deep harsh croak uttered while in flight.

Common Sandpiper

Cattle Egrets

Cormorants

Cattle Egrets

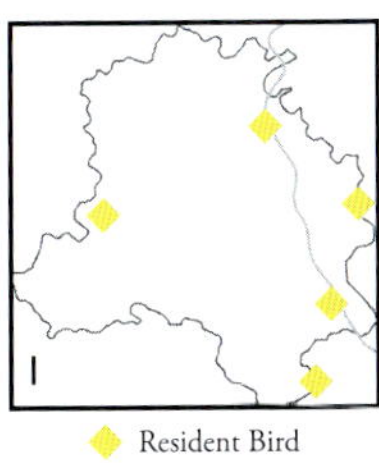

Size:

Cattle Egret - *51 cm*

Intermediate Egret - *80 cm*

Little Egret - *63 cm*

Large Egret - *91 cm*

Hindi name:

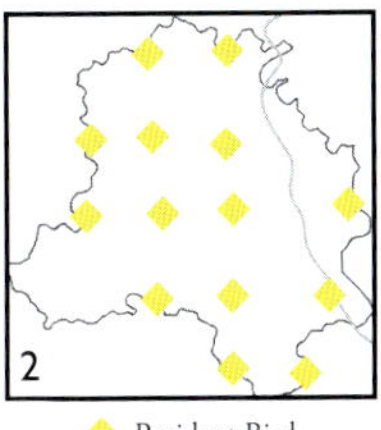

Cattle Egret - *Gai Bagla*

Intermediate Egret - *Karchia Bagla*

Little Egret - *Kilchia Bagla*

Large Egret - *Bada Bagla*

Scientific name:

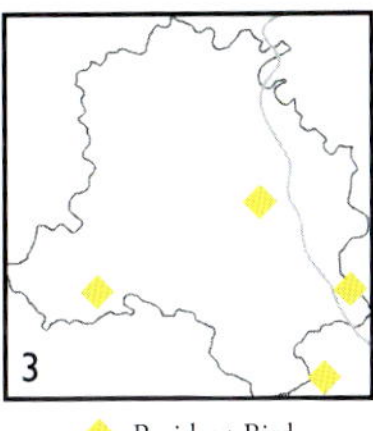

Cattle Egret - *Bubulcus ibis*

Intermediate Egret - *Mesophoyx intermedia*

Little Egret - *Egretta garzetta*

Large Egret - *Casmerodius albus*

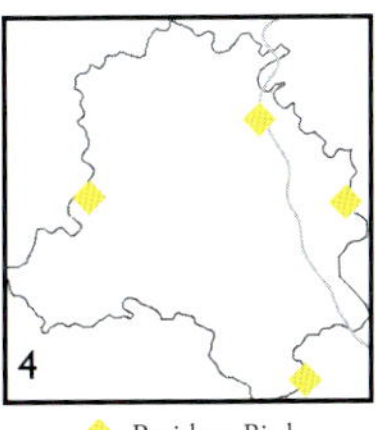

All the four Egrets are pure white in colour. The Cattle Egret is similar in size to the Little Egret and is distinguished only by the colour of its bill which is yellow while in the Little Egret it is black. During the breeding season the cattle egret develops a unique buff-orange head, neck and black-coloured plumage. The Intermediate Egret is slightly smaller but very similar to its large cousin except that it has a shorter bill.

Egrets are spotted in lakes, jheels, rivers, tanks, marshes, etc. You can spot them at the Yamuna and the Okhla Barrage. The Cattle Egret is also spotted near grazing cattle. All except the Large Egret are very gregarious birds.

The nesting season of the Large and Intermediate Egret is between July and February, the Cattle Egret from June to August, and the Little Egret, July and August. The nests of the Large and Intermediate Egret is a fragile platform of sticks on trees near water. The eggs laid are 3 to 4, pale green in colour. Both sexes share the domestic duties.

The nest of the Cattle Egret is an untidy twig platform. It lays 3 to 5, pale blue eggs. The nest of the Little Egret is a twig platform, lined with leaves, straw, etc. The eggs laid are 4 and are a dull bluish-green.

Food - Cattle and Little Egret: grasshoppers, cicadas and other insects; also lizards, fish, frogs etc Intermediate and Large Egret-fish, frogs etc.

Call - Occasionally croaking sounds made by all the Egrets.

GREAT CORMORANTS

COMMON SANDPIPER

DARTER

Mallard

Winter Migratory Bird

Size - *61 cm*

Hindi name - *Nilsir*

Scientific name - *Anas platyrhynchos*

The Mallard is called the 'king of ducks' because it looks so regal in its resplendent colours. The breeding male can be easily identified. It has a green head and a black rear. The males have a yellow bill with a black tip, whereas the females have a dark-brown bill. The females are light brown, with plumage much like most female ducks.

The Mallard is a highly gregarious bird, outside of the breeding season and will form large flocks. It is a bird of most types of wetlands, including parks and small ponds and rivers. It is a swift flier and a noisy bird.

The Mallard breeds only in Kashmir. The nesting season is May and June. Its nest is a pad of rushes, under a bush or a clump of grass near the edge of a lake. It lays 6 to 10, greenish-grey eggs.

Food - Seeds, grass shoots and water plants; also fish, worms, tadpoles, etc.

Call - The male has a nasal loud heep, whereas the female has the familiar 'quack' associated with ducks.

Spot-Billed Duck

Northern Shoveler

Common Coot

Northern Shoveler

Winter Migratory Bird

Size - *51 cm*

Hindi name - *Tidari*

Scientific name - *Anas clypeata*

The Northern Shoveler is easily identifiable due to its large bill which gives it its name. It looks like a spatula or to be more apt a shovel. The breeding male has a green head, a white breast and a chestnut belly. The edges of the wings are chestnut coloured. The females are light brown with a grey forewing. In non-breeding plumage, the male looks more like the female.

The Northern Shoveler is not as gregarious as some other ducks outside the breeding season and forms only small groups. At the Yamuna, you can find many flocks of these quaint-looking and pretty ducks.

The Northern Shoveler is a bird of open wetlands, such as wet grassland or marshes and feeds by using its bill to strain food from the water. This is a somewhat quiet species compared to other ducks.

The Northern Shoveler does not nest in the Delhi region or in India for that matter.

Food - Crabs, snails, slugs, water insects, worms, water weeds and shoots.
Call - Usually silent.

COMMON COOT

SPOT-BILLED DUCK

PHEASANT-TAILED JACANA

Spot-billed Duck

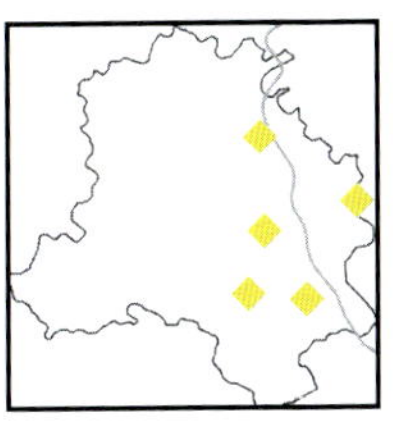
Resident Bird

Size - *60 cm*

Hindi name - *Gugral*

Scientific name - *Anas poecilorhyncha*

The Spot-billed Duck is so named because of the yellow spot on the tip of its bill. Its plumage appears like the light and dark brown scales on a fish. It has two orangish-red spots on the lower portion of its body and orangish-red legs. What a colourful duck! And to think that we always draw white ducks in our drawing class at school.

The Spot-billed Duck is probably the most common and widespread duck of our area. It is a fast and strong flier and is found in pairs or small groups in jheels, ponds, rivers, etc. You can spot it at the Yamuna and at the Delhi Zoo. It generally does not mingle with other ducks.

The nesting season of the Spot-billed Duck is not exactly defined, but mostly it is between July and September. Its nest is a pad of grass and weeds amongst marshy edges. It lays 6 to 12, greyish-buff or greenish-white eggs.

Poor ducks, just because they float on the surface of the water, they are in ancient mythologies sometimes associated with superficiality.

Food - Wild grain, seeds, shoots of water plants and sometimes earthworms and insects.

Call - A loud and noisy *quack-quack.*

Pheasant-Tailed Jacana

Common Coot

Bronze-Winged Jacana

Common Coot

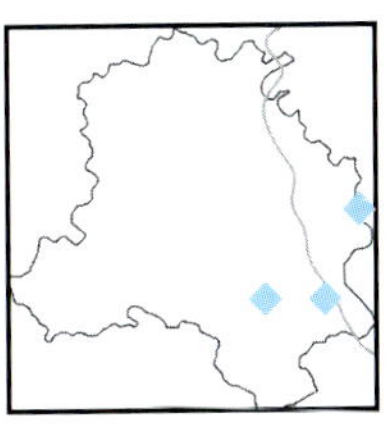

Winter Migratory Bird

Size - *42 cm*

Hindi name - *Dasari*

Scientific name - *Fulica atra*

The Common Coot is a bird one can easily identify from a distance. It is dark grey in colour with a black head, a black neck, a white forehead and a white beak. It has 'bloodshot' red eyes and very large feet. It swims like a duck and is therefore, sometimes misidentified as some breed of duck.

Common Coot sociably gathers in huge flocks during winter, on jheels and tanks. At the Yamuna and at Sultanpur, you can spot hundreds of these birds swimming around. It is a mesmerizing sight to watch these glistening black birds with bewitching red eyes, almost covering the surface of the lake or pond, and the sun sparkling on the water. Only if they have flashed upon the 'inward eye' of William Wordsworth he might has written about them and not the daffodils.

Although Common Coot spends most of its time on water, it is a very swift flier.

The nesting season of the Common Coot is chiefly between July and August. Its nest is a large mass of rushes among matted reeds. It lays 6 to 10 buff-coloured eggs, spotted with reddish-brown or purplish-black.

In 'ye olde English' as she was spoken by the 'propah' Englishmen decades ago, 'coot' meant a stupid, foolish person (you silly old coot!)

Food - Grass and paddy shoots, water weeds, insects, snails, etc.
Call - A clear and loud trumpeting sound.

Bronze-Winged Jacana

Pheasant-Tailed Jacana

White-Breasted Waterhen

Pheasant-tailed Jacana

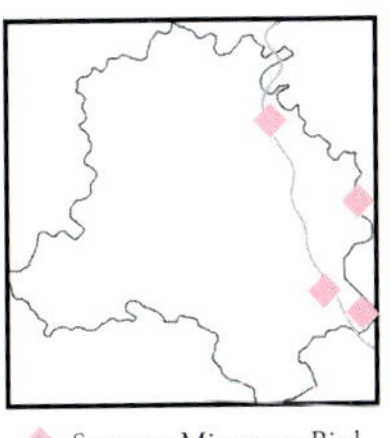

Summer Migratory Bird

Size - *31 cm (excluding tail)*

Hindi name - *Piho*

Scientific name - *Hydrophasianus chirurgus*

The Pheasant-tailed Jacana is a beautiful bird. The breeding adults are black with white wings, a white head and a white foreneck. The back of the neck is golden. The adults have a long pheasant-like tail. There is a white eyestripe and the legs and toes are greyish. Non-breeding adults don't have the long tail. The belly is white except for a brown band on the breast and a neck stripe. The side of the neck is golden. The toes are unusually long.

The Pheasant-tailed Jacana is found alone or in scattered groups on vegetation covered lakes and jheels.

The nesting season of the Pheasant-tailed Jacana is between June and September. Its nest is a skimpy pad of weed-stems, etc, on floating leaves often partially submerged. It lays 4, glossy, greenish-bronze or rufous-brown eggs. The female is polyandrous, which means it can mate with more than one male. That's one super liberated bird!

Food - Vegetable matter, insects, snails, etc gathered from the floating vegetation on the water surface.

Call - A unique nasal *tewn tewn* type of mewing sound.

WHITE-BREASTED WATERHEN

BRONZE-WINGED JACANA

COMMON GREENSHANK

Bronze-winged Jacana

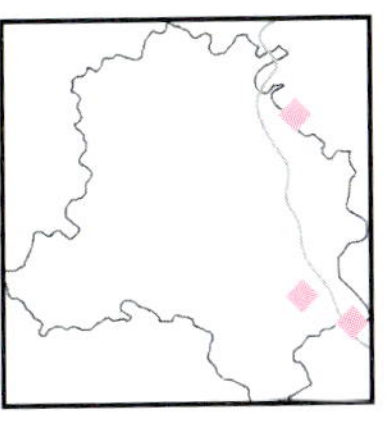

Summer Migratory Bird

Size - *28 cm*

Hindi name - *Dal pipi*

Scientific name - *Metopidius indicus*

Like the Pheasant-tailed Jacana, the Bronze-winged Jacana is a beauty. It is largely black, although the inner wings are a chocolate brown and the tail is red. There is a prominent white eyestripe. The bill is yellow. The female is larger than the male.

The Bronze-winged Jacana is found mainly on vegetation covered lakes, ponds and jheels. In Delhi however, it is the rarer of the two Jacanas. The Bronze-winged Jacana is a very good swimmer.

The nesting season of the Bronze-winged Jacana is between June to September. Its nest is a pad of weed-stems, etc, floating leaves often partially submerged. It lays 4 lovely glossy eggs.

It's a beautiful sight to watch this delicate bird elegantly walking in shallow water, daintily lifting its slender legs, very much like a ballerina.

Food - Seeds, roots of water plants and insects, snails, etc, picked from the vegetation on the water surface.

Call - A short hoarse grunt and a wheezy *seek-eek-seek*.

Common Greenshank

White-Breasted Waterhen

Common Redshank

White-breasted Waterhen

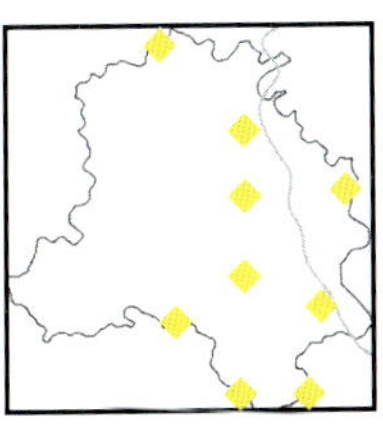

Resident Bird

Size - *32 cm*

Hindi name - *Dawak*

Scientific name - *Amaurornis phoenicurus*

The White-breasted Waterhen is a grey, long-legged and short-tailed marsh bird. It has a white head, a white breast and a red patch under the tail. It is seen alone or in pairs, near bushes, thickets, reeds and on marshy surfaces. It jerks its stubby tail, especially while moving.

The White-breasted Waterhen is mostly a shy and quiet bird but becomes very noisy during the monsoon season when it breeds. It keeps silent during the rest of the year. You can spot it near jheels and marshes.

The nesting season of the White-breasted Waterhen is between June and October. Its nest is a shallow cup-shaped structure of twigs and stems, in a bush near water. It lays 6 to 7, creamish or pinkish-white eggs, streaked with reddish-brown.

It is an amazing fact that the White-breasted Waterhen can climb trees! Its quite a sight, if you are lucky enough to be a witness to this simian feat of an avian creature!

Food - Insects, snails, worms, grain and shoots of paddy and marsh plants.
Call - It starts off with noisy, hoarse grunts and croaks then changes to a monotonous *krr-kwaak-kwaak* etc.

COMMON REDSHANK

COMMON GREENSHANK

BLACK-WINGED STILT

Common Greenshank

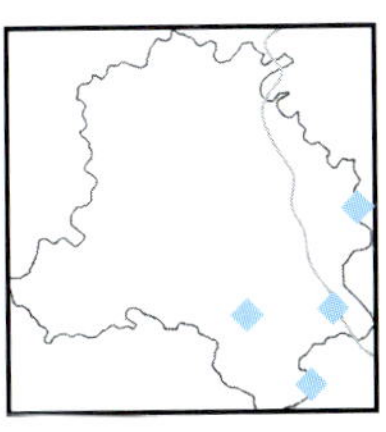

Winter Migratory Bird

Size - *36 cm*

Hindi name - *Tantana*

Scientific name - *Tringa nebularia*

The Common Greenshank is greyish-brown above with a whitish head and belly. It is streake with dark marks on the upper parts. The legs are green. It has a black rather upturned beak.

The Common Greenshank is seen alone or in small groups often with redshanks and other waders. It is usually spotted at the edges of water bodies like lakes, swamps, rivers etc. In Delhi, you can spot them on the banks of the Yamuna.

The Common Greenshank is a very suspicious bird and takes off at the slightest noise or movement. It is a fine flier.

The Common Greenshank nests in Northern Europe and Asia up to the Arctic circle, and not India.

Food - Insects, tadpoles, frogs, crabs, snails, slugs, etc.
Call - Similar to the redshank but of lower pitch.

Black-Winged Stilt

Common Redshank

White-Breasted Kingfisher

Common Sandpiper

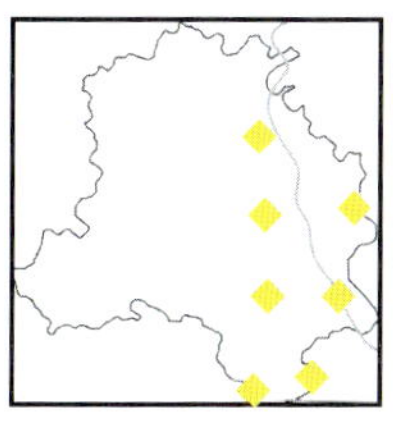

Size - *21 cm*

Hindi name - *not recorded*

Scientific name - *Actitis hypoleucos*

The Adult Common Sandpiper has short, yellowish legs and a pale lower bill with a dark tip. The body is greyish-brown above and white underneath. It has dark streaks on its foreneck. It has a unique stiff-winged flight and flies low over water.

The Common Sandpiper, is actually not that common. It is not a gregarious bird, rarely seen in large flocks, it is spotted singly at ponds, tanks, lakes etc. Look for them at the Yamuna and also at the Delhi Zoo. These birds hunt for food on the ground or in water.

The Common Sandpiper does not nest in the Delhi area, restricting itself to the hills of Kashmir, Kumaon, etc. Its nesting season is may and June.

Food - Insects, crabs, shrimps, etc.

Call - A sharp tee-tee-tee, also a shrill wheeit...wheeit repetitive sound when disturbed.

Butterflies

Common Bluebottle

Blue Pansy

Common Castor

Blue Pansy

Scientific name - *Junonia orithya*

Wingspan - *45-60 mm*

The Blue Pansy is a gorgeous-looking butterfly. It reminds you of Keats' immortal line, 'A thing of beauty is a joy forever.' This epitome of beauty never fails to fill me with joy. The hindwings are a beautiful blue, and the forewings are brown and black with two prominent white bands.

The Blue Pansy can be spotted in the dry and open areas of Delhi. This butterfly is the most active during the warm hours of the day. It is a sun-loving insect, which can usually be found on grassy patches in open areas. The Blue Pansy is an easily approachable butterfly and does not really mind if you get a bit 'upclose and personal'. So please don't break its trust by catching hold of it. It can fly quite speedily and occasionally engages in a spiral dance going up some distance from the ground.

I am really lucky to have been treated to the ethereal vision of a Blue Pansy doing its spiral number, right across my house in the District Park. She reminded me ofTinker Bell!

Common Castor

Common Bluebottle

Common Cerulean

Common Bluebottle

Scientific name - *Graphium sapedon*

Wingspan - *80-90 mm*

The Common Bluebottle is a brilliant-looking butterfly. It has a bluish-green band which runs from the apex of the forewing to the internal margin of the hindwing on both the upper and the underside of its wings. The hindwing has a chain of blue spots on the upperside, and an additional red spot on the underside of the hindwing. There is a red spot near the base of the hindwing on the underside as well.

The Common Bluebottle male can often be found feeding on roadside seepage. This quick-flying butterfly is common in nature reserves like the Sultanpur Park. In flight, you can catch a glance of its brilliant blue wings. Females are more difficult to spot than the males.

Common Cerulean

Common Castor

Common Crow

Common Castor

Scientific name - *Ariadne merione*

Wingspan - *45-60 mm*

The Common Castor is also known as the Castor Butterfly. It is a lovely rust-coloured butterfly. It has wavy black lines on its wings. This looks like a delicate zigzag lace pattern on a piece of rust coloured silk! The underside of the wings is much darker, but the colour is rust.

The Common Castor is spotted in dry and grassy areas. It is also seen near sewage drains and water run-offs. It is a lazy butterfly, 'loitering' around casually! Its flight consists mainly of glides and is short. The movements appear quite ungainly and clumsy. It either gets tired quickly and rests, or does not like others to see its 'gawky gait' so cuts short its flight.

The Common Castor is called so because its larvae feed almost exclusively on the Castor oil plant.

Common Crow

Common Cerulean

Common Grass Yellow

Common Cerulean

Scientific name - *Jamides celeno*

Wingspan - *27-40 mm*

The Common Cerulean is an attractive-looking butterfly. The underside of the wings is greyish-white during the rainy season and dull brown in the dry season. It is marked with straight lines. It also has a tail. Both the male and the female have glossy blue uppersides. The male is brighter than the female and has narrower dark borders on its wings.

The Common Cerulean is a familiar butterfly. It is often seen near human habitation, well-wooded areas and near water. It likes partially sunlit areas and can be spotted lazing around in the shade or in soft sunlight. It never basks with its wings spread out open. Its flight is uncertain and not fast. It does not like to alight on the ground and prefers to flit around amongst bushes or sometimes small trees and shrubs.

The Common Cerulean feasts on the nectar of flowers like, Tridax, Sida, Chromolaena, etc. The males also feed on wet soil.

Common Grass Yellow

Common Crow

Common Gull

Common Crow

Scientific name - *Euploea core*

Wingspan - *85-95 mm*

The Common Crow Butterfly unlike the avian crow, is so elegant to look at, that if you spot one you will not be able to take your eyes off it! The Common Crow is a shiny black butterfly with brown undersides. It has white spots on the outer wings, as well as on the body. The antennae, head, thorax and abdomen are dark brown.

The Common Crow is common all over Delhi. It belongs to the Danaid subfamily of the Nymphalidae family. This beautiful-looking creature is very 'bitter' inside because as a caterpillar it feeds on the latex of food plants. The chemicals in this latex make this butterfly a very distasteful meal. It has a tough skin to survive an occasional attack. When attacked, it oozes out a foul liquid which smells and tastes terrible. This forces the attacker to release it quickly. It recovers mysteriously as soon as it is free of its enemy, and flits around in a leisurely manner, gliding elegantly.

Common Gull

Common Grass Yellow

Common Evening Brown

Common Grass Yellow

Scientific name - *Eurema hecabe*

Wingspan - *40-50 mm*

The Common Grass Yellow is a very delicate butterfly. The wings are bright yellow with the tip and the outer edge generally black. The forewing has two black spots (one or both of which may be missing) on the under surface. The upper surface of the rear wing has a narrow black border like a 'delicate piping on a dress'.

The Common Grass Yellow belongs to the Pierids family of Butterflies. It is a very common butterfly of Delhi and is found throughout the region. The Common Grass Yellow's food is Cassina and other leguminous plants. Though its flight is rapid, it flies in an irregular manner. It is a fearless little creature which flits around through all seasons except may be when it is exceptionally cold.

Common Evening Brown

Common Gull

Common Emigrant

Common Gull

Scientific name - *Cepora nerissa*

Wingspan - *40-65 mm*

The Common Gull is a fine specimen. It is white overall on the upperside and yellow on the underside. The wing margins and the veins are black. The females have more prominent markings. During the monsoon it turns into a lovely yellow and deep black. A little taxi on wings?

The Common Gull is a common butterfly in the scrub and dry areas of the Delhi region. It also enters gardens and parks. It flies over small trees, bushes and grass, looking for nectar. Flower nectar is its main source of food. The Common Gull has a swift flight and always remains close to the ground. It is most active after sunrise uptil late afternoon.

I have a specially soft corner for this butterfly because its a 'gull' and the word reminds me of my favourite book, my bible almost, Jonathan Livingstone Seagull

Common Emigrant

Common Evening Brown

Common Pierrot

Common Evening Brown

Scientific name - *Melanitis leda*

Wingspan - *60-80 mm*

The Common Evening Brown is an inconspicuous brown butterfly. The upperside is a dark brown with an eye-spot and a white 'pupil' like mark on the forewings surrounded by orange patches. The underside has varying colours. The Common Evening Brown rests on fallen dry leaves and camouflages with them. So, it is very difficult to spot.

The Common Evening Brown adapts to many different areas such as dense forests, open areas, scrubland, gardens etc. As its name suggests, it is most active during the evening hours. It visits flowers of Ixora, Paveta and Impatiens. It also likes fallen and rotten fruits, and can be often seen feeding on them. The Common Evening Brown keeps close to the ground and never flies very high up. While on the ground, it pins up its wings and displays its eye-spots. This appears as if its looking at you!

Common Pierrot

Common Emigrant

Common Mormon

Common Emigrant

Scientific name - *Catopsilla pomona*

Wingspan - *55-80 mm*

The Common Emigrant looks like a petal of a yellow rose. The colour of the wing ranges from white with yellow markings, to completely plain yellow.

The Common Emigrant is a very common and a variable species of butterfly. It is a fast flier, covering long distances high above the ground in a straight, powerful, long, and curved flight. It is more easily spotted in wet places or during the rainy season. Common Emigrants visits flowers of ceylon carissa, lantana, fiddle-leaved jatropha, poinsettia, bougainvillea, pomegranate, cotton, sunflower, etc.

This butterfly gets its name from its habit of migration. However unlike birds, many insects migrate in one direction and may not go back to the please from where they have migrated.

Common Mormon

Common Pierrot

Common Wanderer

Common Pierrot

Scientific name - *Castalius rosimon*

Wingspan - *24-32 mm*

The Common Pierrot is a petite and tiny butterfly. It is white overall, but the underside of the wings has black spots. The lower corner of the wing has a shiny green spot. It has a short, white-tipped black tail. The bases of the wings on the upperside are green. The borders of the wings are brownish-black with small white spots. Both the male and female are similar in appearance.

The Common Pierrot can be spotted in scrub areas, grasslands and also near human settlements. It has a feeble flight, and always flies close to the ground repeatedly landing on leaves. Likes to keep safe, this little fairy. It is fond of sunlight and basks more after the rains. The Common Pierrot feeds on flowers like Sida and Tridax, wet soil, dead insects and even bird droppings.

Common Wanderer

Common Mormon

Danaid Eggfly

Common Mormon

Scientific name - *Papilio Polytes Polytes*

Wingspan - *90-100 mm*

The Common Mormon belongs to the swallowtail family. It is an impressive-looking butterfly. The male is a rich-black with a series of white spots in the centre of the hindwing. The edge of the forewing too has a row of white spots. The female occurs in three different colour forms, one resembling the male and the other two mimicking the Common Rose and the Crimson Rose butterflies. The males are smaller in size.

The Common Mormon is the most common swallowtail butterfly. It is found near habitations chiefly in gardens and fields which have citrus fruit trees like oranges, lemons, etc. It can also be spotted in open woods and gardens at low altitudes. It flies close to the ground around bushes and, hedges but stays away from wet areas. The male is a faster flier than the female.

The Common Mormon visits the flowers of dividivi, acacias, cassia, etc. It feeds on cultivated citrus, especially lime berry, orange, lime, curry leaf, bael, garden rue, etc.

Danaid Eggfly

Common Wanderer

Grass Jewel

Common Wanderer

Scientific name - *Pareronia valeria*

Wingspan - *65-80 mm*

The Common Wanderer presents a pretty picture. The male is a glossy light blue overall with black edges to the veins. The female is a light-bluish white with broader edges and veins. The markings on the underside are similar to the ones on the upperside.

The Common Wanderer is found in open forests, scrub areas and woodlands, gardens etc. Its typical wandering flight gives it its name. The male flies tirelessly, rarely resting. The flight is moderate, but if disturbed it flies quite quickly. The male is more commonly seen than the female, since it mimics the Blue Tiger butterfly.

The Common Wanderer feeds on flowers like the plumbago, which have longer corolla tubes.

Grass Jewel

Danaid Eggfly

Indian Cabbage White

Danaid Eggfly

Scientific name - *Hypolimnas misippus*

Wingspan - *70-85 mm*

The Danaid Eggfly is also known as the Common Eggfly and the Diadem. The female is yellowish-brown. The apex of the forewing is black with a white band. The underside is similar but a bit duller. The male is black above, and has two shiny white spots, a large one on the hind wing and a small spot on the forewing. The underside is a pale rusty-brown. A wide white band on the hind wing and a thin one on the forewing stand out.

The Danaid Eggfly is an inhabitant of open country with moderate rainfall. It is a strong and fast flier. The male is more common than the female. The male is highly territorial and is often found basking in the sun by resting on bushes or on the ground with a continuous slow movement of wings to the sides.

The Danaid Eggfly visits flowers of lantana, common zinnia, ceylon carissa, sunflower, aztec marigold, golden groundsel, vipersbugloss, etc.

The Danaid Eggfly female is an outstanding mimic of the Plain Tiger The Plain Tiger is a bitter-tasting butterfly, and so is left alone by predators. Therefore, to fool its enemy, the female Danaid Eggfly performs this intelligent act of aping the Plain Tiger's appearance!

Indian Cabbage White

Grass Jewel

Lime Butterfly

Grass Jewel

Scientific name - *Freyeria Trochylus*

Wingspan - *15-22 mm*

The Grass Jewel is really a precious jewel found amongst grass. It is the smallest butterfly in India. It looks like a small piece of paper flying around, being carried around by the breeze! The males have a brown upperside. Specimens found in dry areas are comparatively paler than their brethren from areas of heavy rainfall. The underside is a pale silky brown and spotted with light brown. There are metallic orange spots at the edges of the wings. The females are brown.

This jewel of grass would be a befitting jewel for any woman wishing to be seen in all her finery.

Lime Butterfly

Indian Cabbage White

Peacock Pansy

Indian Cabbage White

Scientific name - *Pieris canidia*

Wingspan - *45-60 mm*

The Indian Cabbage White is a cute little butterfly. The apex of the forewing on the upperside is black with some black spots. The hindwings too have black marginal spots. There is a big black spot on the outer forewing. The underside of the hindwings is dusty white, with dull yellow and grey scales. The female has an additional black spot on the upperside of the forewings.

The Indian Cabbage White avoides dense forests and keeps to shrubs and bushes. As its name suggests, it is a cabbage pest. Although, not a major threat, it sometimes poses a serious problem when the population of this butterfly increases rapidly and attacks crops.

The Indian Cabbage White is the most active around afternoon. It has a slow and fluttering flight, and usually flies low.

The Indian Cabbage White feeds on flowers of herbs and shrubs like Impatiens, Crepis, Vicoa, etc. The wings are closed while it feeds on flowers.

Peacock Pansy

Lime Butterfly

Plain Tiger

Lime Butterfly

Scientific name - *Papilio demoleus*

Wingspan - *80-100 mm*

The Lime Butterfly belongs to the swallowtail family. It is one of the most common butterflies of our area. It has black wings that turn brown with age. The wings are spotted with yellow. On the underside it is a pretty, yellow, red and blue. The male and female are similar in appearance. It doesn't have the characteristic tail of other Swallowtail butterflies.

The Lime Butterfly basks with its wings held wide open on tufts of grass, herbs and generally keeps close to the ground, even on cloudy days. It is an interesting butterfly in that it has a number of ways of flying. In the cold of the morning the flight is slow; as the day progresses, it flies fast, straight and low. In the hotter part of the day it may be found settling on wet soil where it will remain still, if left alone except for an occasional flutter of wings.

The Lime Butterfly is a regular visitor of flowers and shows a preference for flowers of smaller herbs rather than larger plants such as the lantana. While resting, it closes its wing over its back and draws the forewings between the hindwings.

PLAIN TIGER

PEACOCK PANSY

PIONEER

Peacock Pansy

Scientific name - *Junonia almana*

Wingspan - *60-65 mm*

The Peacock Pansy is as pretty as a peacock! Should it be honoured with the title, 'National Butterfly of India'? It is bright and tawny, and has large eye-spots on the wings. The undersides also develop eye-spots during the rainy season. In the dry season the underside resembles a leaf.

The Peacock Pansy is found relaxing during early morning hours in sunny patches, often surrounded by darker shady areas. At this moment it appears like an Orange-face Owl Butterfly, thus scaring away likely invertebrate predators. It is highly aggressive in behaviour and often chases away intruders!

The Peacock Pansy likes to visit flowers such as the marigold, etc.

Pioneer

Plain Tiger

Salmon Arab

Plain Tiger

Scientific name - *Danaus Chrysippus*

Wingspan - *70-80 mm*

The Plain Tiger is a beautiful butterfly. It is medium-sized. The male Plain Tiger is smaller than the female. The body is black with many white spots. The wings are orangish-brown, the upper side being brighter and richer than the underside which is a little dull. The hind wing has three black spots around the centre.

The Plain Tiger, is commonly found in India. It belongs to the Danaid subfamily of the Nymphalidae Family, that is, the Brushfooted butterflies.

The Plain Tiger is mainly a butterfly of open country, parks and gardens. This butterfly is perhaps the commonest of all Indian butterflies and almost everyone must have sighted it sometime or the other. It flies around the whole day, sucking nectar from flowers and as the sun begins to set, this tiny fairy-like creature skims over bushes to find a resting place for the night.

Its amazing how nature never fumbles in its patterns. Every Plain Tiger Butterfly born on this earth has had three and only three black spots!

Salmon Arab

Pioneer

Striped Tiger

Mehran Zaidi

Pioneer

Scientific name - *Anaphaeis aurota*

Wingspan - *40-55 mm*

The Pioneer is a common and pretty butterfly. The upperside is white with black markings. The hindwings are unmarked except at the wing edges. The underside is a bright yellow with black bands along the veins. The males are brighter and have thinner black bands. The females are normally larger in size.

The Pioneer is a butterfly which prefers dry and scrub areas. It has a swift flight, with fluttering wing movements. The Pioneer likes to feed on flowers like lantana, gmelina, tridax etc. The Pioneer is a very alert creature and takes off at the slightest sign of danger. While resting, it shows signs of laziness and that is the time to watch this lovely little winged creature.

I often wonder what this tiny creature must have pioneered to be called The Pioneer.

Striped Tiger

Salmon Arab

Yellow Orange Tip

Salmon Arab

Scientific name - *Colotis amata*

Wingspan - *35-40 mm*

The Salmon Arab is a small black and salmon-coloured butterfly. The broad black bands on the upper side are spotted with a variable number of light-coloured spots. The underside of both sexes is powdered with grey scales. The male is always salmon coloured above and yellow below, with a light dusting of grey scales.

The female has two colour forms. The commoner form resembles the male but is of a lighter shade of salmon-pink and has a well marked underside. In the other form, the salmon-coloured areas are substituted by dull yellow to creamish-white, and the underside is lightly marked. The dry season form is smaller and lighter in colour than the wet season form.

The Salmon Arab can be spotted in scrub and thorny areas. It is a very vivacious butterfly and may be seen on the wing even during the hottest part of the day. It flies low to the ground and flies rapidly with a constant flapping of its wings, repeatedly settling on flowers. Once settled, it holds its wings up at an angle and often tilts them to cover the hind wing.

All the Arabs I have seen are always dressed in white, so why is this 'Arab' salmon? In fact why is it called an Arab?

Yellow Orange Tip

Striped Tiger

Yellow Pansy

Striped Tiger

Scientific name - *Danaus genutia*

Wingspan - *72-100 mm*

The Striped Tiger bears a close resemblance to the Royal Bengal Tiger. But that is where the comparison ends! This is a gentle butterfly a prey rather than a predator. The wings are tawny with broad black stripes. The wing margins are black with two series of white spots. The underside is similar in colour but is duller. The male has a black-and-white spots on the hindwing.

The Striped Tiger is a common butterfly frequenting scrub areas, and land around human settlements. It is a strong flier, but is not a fast one.

The striped Tiger feeds from the flowers of lantana, celosia, cosmos, zinnia and others. While sleeping its wings are closed over its back.

If the Royal Bengal Tiger continues to be an endangered cat and gradually becames extinct this 'tiger' might be all we would have to remind us of the royal feline.

Yellow Pansy

Yellow Orange Tip

White Orange Tip

Yellow Orange Tip

Scientific name - *Ixias pyrene*

Wingspan - *50-70 mm*

The Yellow Orange Tip looks like a lovely flower a marigold or a chrysanthemum perhaps. It is a pretty yellow butterfly with black and orange wing margins.

The Yellow Orange Tip flies more or less in a straight line and relaxes after a long flight on leaves in bushes and trees. During dull weather or in the evenings, it rests at places lower down among leaves in the undergrowth. That's a pretty safe 'bedroom' for this delicate creature.

The Yellow Orange Tip visits flowers of lantana, ber, batchelor's button, aztec marigold, pomegranate, chrysanthemum, horse-purslane, indian abutilon, etc.

WHITE ORANGE TIP

YELLOW PANSY

BLUE PANSY

Yellow Pansy

Scientific name - *Precis hierta*

Wingspan - *45-60 mm*

The Yellow Pansy is an exqusitie-looking butterfly, like its cousin the Blue Pansy. It is largely yellow, the forewing apex is black with yellow spots. The hindwings have a large blue spot each.

The Yellow Pansy is found in the drier areas and near stony stream-beds. It can be seen basking in the sun on the ground or on rocks. The Yellow Pansy likes to visit flowers for food. Its favourite is the marigold.

The Yellow Pansy belongs to the Nymphalidae family of butterflies which has about 5,000 species! One illustrious 'scion' of this family is the Monarch Butterfly.

Blue Pansy

White Orange Tip

Common Bluebottle

White Orange Tip

Scientific name - *Ixias marianne*

Wingspan - *50-55 mm*

The White Orange Tip is truly a great sight. It is white on the upperside with broad black borders on the wings, and a beautiful orange forewing apex. The female has a series of black spots on the portion where the male has orange ones. The underside is glossy yellow with some brown specks on the hind wings.

Like its other Pierid cousins, the white orange tip too is a butterfly of dry and scrub areas. It likes to fly low among bushes and small trees, looking for flowers. Its flight is swift and its wing-beats are continuous. While it feeds, it draws its forewings in between its hindwings.

I sign off with a little prayer for butterflies.

May the Lord keep you, ye gentle creatures,
ye corporeal angels.
And may your dazzling beauty always bring joy
to the tried eyes of cynical Homo Sapiens.